£9.95

BASIC fluid mechanics

BASIC fluid mechanics

J J Sharp BSc, MSc, ARCST, PhD, FICE, FCSCE, PEng
Faculty of Engineering, Memorial University, Newfoundland

Butterworths
London Boston Durban Singapore Sydney Toronto Wellington

First published 1988

© **Butterworth & Co. (Publishers) Ltd, 1988**

British Library Cataloguing in Publication Data
Sharp, James J.
  Basic fluid mechanics.—(Butterworths
BASIC series).
  1. Fluid mechanics—Data processing
  2. BASIC (Computer program language)
  I. Title
  532'.0028'55133   TA357

  ISBN 0-408-01640-X

Library of Congress Cataloging-in-Publication Data
Sharp, J. J. (James J.)
  BASIC fluid mechanics.

  (Butterworths BASIC series)
  Includes index.
  1. Fluid mechanics—Data processing. 2. BASIC
(Computer program language) I. Title. II. Series.
  TA357.S443   1987   620.1'06   87-23921
  ISBN 0-408-01640-X

Photoset by August Filmsetting, Haydock, St Helens, Lancs
Printed and bound by Page Bros Ltd, Norwich, Norfolk

Preface

In common with other texts in the series this book combines the application of BASIC programming with an engineering discipline, namely fluid mechanics. Only a brief introduction is given to the use and application of BASIC (Chapter 1) but this is sufficient to permit a beginner to understand and to write BASIC programs. Many excellent, well written books providing full details of the language are available and some of these are listed in the bibliography. Following chapters deal with various topics in fluid mechanics. These are explained in some depth and worked examples developing programs for the solution of typical problems are provided. Once again sufficient information is provided to permit the student to gain considerable understanding of the subject. All the topics considered would be suitable for a general introductory undergraduate course in fluid mechanics given to students in any engineering discipline. More advanced topics in civil engineering are provided in *BASIC Hydraulics* and *BASIC Hydrology*. Taken together these three texts provide a comprehensive introduction to water engineering.

All the programs in the text (with the exception of the matrix program in Chapter 9) were written in GW BASIC and run on a Sanyo 885 personal computer. These will run on any IBM-compatible machine. None of the programs requires a printer and all provide output to a monitor screen. Relatively little memory is required.

Although the coverage of BASIC programming and fluid mechanics is not as exhaustive as would be found in specialty texts, sufficient explanation of both is given to permit the text to be used by someone with no knowledge of either. The book should therefore be found useful by students in an undergraduate program or by practising engineers who are attempting to get to grips with modern computational procedures. The programs developed in the worked examples have been deliberately kept simple so that they may be easily understood and reproduced with the minimum of trouble. In many cases, however, the basic procedures used can be developed to provide much more sophisticated programs such as those which might be required for practical work.

J J Sharp

To
Glenn, Jennifer and Michael

Contents

Chapter 1

Introduction to BASIC

1.1 Introduction

BASIC is an acronym for 'Beginners All Purpose Symbolic Instruction Code'. It was developed at Dartmouth college in the 1960s by Professors Kemeny and Kurtz for use on a time-sharing system. The language is user oriented and makes uses of instructions resembling basic algebraic formulae augmented by certain easily understood words such as LET, GO, TO, READ, PRINT, IF, THEN, etc. With the advent of the small relatively cheap personal computer, BASIC is now one of the most popular and most widely used of all computer languages. Other popular time-sharing languages include ALGOL, COBOL, FORTRAN and PASCALL. More specialized languages also exist. However, of all these, BASIC is the easiest to learn and use.

A BASIC program consists of a series of statements or lines, each beginning with a line number followed by a BASIC command. Except when instructed to do otherwise, the computer executes each line in order, beginning at the smallest line number and proceeding to the largest number at the end of the program. Statements do not have to be typed in order when the program is written. Figure 1.1 illustrates a simple BASIC program designed to calculate the area of a

```
100 INPUT R
200 LET A=3.142*R*R
300 PRINT A,R
400 END                 Figure 1.1
```

circle. The program consists of four statements and includes the BASIC keywords, INPUT, LET, PRINT, END. The first statement (100) allows the operator to specify the radius of the circle. The second statement (200) calculates the area A. The third statement prints the area and the radius, and the last statement identifies the end of the program. The key words used in this simple program, and other key words which permit alternative operations, will be described later in more detail.

Various steps must be followed in solving any problem. It is important that the problem be clearly defined and that the user should understand the problem completely. Following definition, the prob-

1

lem must be described in mathematical terms and a formula or algebraic procedure, i.e. an algorithm, must be developed. When this has been done, a flowchart should be drawn to illustrate symbolically the logic of the solution. Then, following these three steps, the program itself is developed using a flow chart, or pseudo code, as a guide to the actual coding. To solve a specific problem the program must then be run on a computer. The first few runs should be used for checking purposes to ensure that the program did in fact do what was intended. This check is most important because it is easy to make errors in syntax or, in a complicated program, to have logic errors present. Finally, when the program is working properly, it is useful to develop some documentation describing it. The necessity for this documentation may not be immediately obvious but, if the program is stored for long periods between use it is very easy to forget the details required for satisfactory operaton. Documentation within the program itself is most important and sufficient REMARK statements should be included to clearly describe the various parts of the program.

1.2 Flow charts

The development of a flow chart assists the programmer to clarify the logic of the program. It specifies the operations which must be carried out by the computer and lists these in order. For example a flow chart to determine the area of a circle based on a specified diameter might take the form shown in Figure 1.2. If several calculations were to be performed using different diameters the flow chart could be modified with a test to determine whether the program should be stopped or re-executed with a different diameter. Figure 1.3 shows such a flow chart developed to determine the volume of a cylinder of length L and diameter D. One simple way of determing when to stop calculations might be to specify zero diameter and to stop the program if the volume is equal to zero but to continue it for all non-zero cases. Such a program is shown in Figure 1.4. Statement 60 transfers control to line 100 when the volume is zero. If the volume is not zero the program continues to line 70 which transfers control back to the beginning and requests the user to specify another value of D. To end the series of calculations the user would at this time specify a zero value for D, or L, or both. Flow charts may be drawn on paper of any size but it is important that the paper be large enough for clarity.

Pseudo code is often used as an alternative to a flow chart. This is neither a computer language nor standard English and might conveniently be thought of as a sort of 'pidgin computerese' which

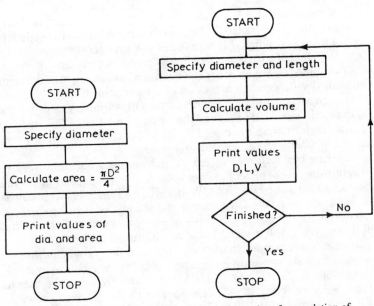

Figure 1.2 Flow chart for calculation of area

Figure 1.3 Flow chart for caculation of volume

```
20 INPUT D
30 INPUT L
40 LET V=3.142*D*D*L/4.0
50 PRINT V,D,L
60 IF V=0 THEN GOTO 100
70 GOTO 20
100 END
```

Figure 1.4

describes the various components of the algorithm in simplistic fashion using words instead of the boxes associated with the flow chart.

1.3 Variables, arrays and strings

Numerical quantities in BASIC are referred to as numbers or constants. They can be represented as integers (whole numbers with no decimal) or real constants which always include a decimal and may or may not have a fractional part. Very large or very small real constants may be expressed with an exponent. For example, the number 7521.3 may be expressed as 7.5213×10^3 and in BASIC would be written as 7.5213E3. Typically, constants may range in value from 10^{-38} to 10^{38}, although the limits vary from one form of BASIC to another. The exponent must be an integer and may be positive or negative.

A numeric variable is a name used to represent a number. Early forms of BASIC restricted the length of simple variables to two letters or one letter and a digit. Modern forms however permit more characters. GW-BASIC, for example allows up to 40 characters. In Figure 1.4 the letters D, L, and V are the variables. These can take different numerical values during execution of the program.

It is important to know the limitations on variable names for the system to be used. IBM PCs and the Apple Macintosh accept up to 40 characters for variable names. The TRS 80 and the Apple II however ignore all characters except the first two.

A string is a sequence of letters and numbers and may vary in length from 15 characters to over 4000 characters depending on the particular version of BASIC. GW-BASIC, for example, accepts up to 255 characters enclosed in double quotation marks. A string variable, used to represent a string, consists of a letter, or series of characters, followed by a dollar sign. These are useful in including text in a program. For example, the test of whether a program is finished or whether calculations must be repeated, could be undertaken by printing a question 'ARE YOU FINISHED?' and requiring the user to type 'YES' or 'NO'. If this were applied to the problem shown in Figure 1.3, lines 60 and 70 of Figure 1.4 would be replaced by the four lines shown in Figure 1.5. Line 60 asks the user if he is finished. Line 70 requests the entry of a string variable, and line 80 tests to see if that string variable is 'YES', in which case the program is stopped.

The rules for incoming numeric variable names also apply to string variable names.

A list, or table, of values is referred to as an array. For example, the area of a circle may be required for 10 different values of diameter. These diameters could then be specified as D1, D2, ... D10. The array in this case is one-dimensional and is simply a list of the diameters. Any particular item on the list can be specified by its number, and this can be represented in a BASIC program by an integer variable. For example, to undertake the calculations of circle area using ten different radii, Figure 1.1 could be modified as shown in Figure 1.6. Lines 100 to 400 remain essentially the same but are written now as arrays. Line 80 sets up a counter and varies the integer I from 1 to 10 so that the calculation is repeated ten times. The loop for calculation is set up between lines 80 and 350. Line 50 is a dimension statement used to reserve the appropriate amount of space. Many formulations of BASIC permit arrays with up to ten items without a dimension state-

```
60 PRINT "ARE YOU FINISHED ? YES/NO"
70 INPUT B$
80 IF B$="YES" THEN GOTO 100
90 GOTO 20
```

Figure 1.5

ment. Two dimensional arrays are used to handle tabulated data. Tables consist of horizontal rows and vertical columns. In specifying a two dimensional array, the variable must be defined by two integers. For example, the variable $X(I, J)$ represents the value stored in row I and column J.

1.4 Input, output and format

Figures 1.1 to 1.6 demonstrated one method of entering data into the program. This was the INPUT statement which permits data to be

```
50 DIM R(10),A(10)
80 FOR I=1 TO 10
100      INPUT R(I)
200      LET A(I)=3.142*R(I)*R(I)
300      PRINT A(I),R(I)
350 NEXT I
400 END
```

Figure 1.6

entered while the program is being run. The most general form is that shown in Figure 1.1, line 100. The statement consists of the line number and the word INPUT, followed by the variable, R, which represents the number to be entered by the operator. When the statement is executed the computer types a question mark and waits until the user has typed in the value which is to be assigned to the variable R.

An alternative method of entering numeric values is to make use of the READ statement together with a DATA statement. The READ statement assigns a specific numeric value to a simple or array variable or assigns a string to a string variable. The values or string to be assigned are contained in a DATA statement. For example the statements.

```
10 READ X,Y,Z

20 DATA 40,32.5,236.1
```

would cause the variable X to be assigned a numeric value of 40. The variable Y would be given a value 32.5 and the variable Z would be set equal to 236.1. DATA statements must be typed within the program but may be situated at any location. Notice that the order in which the data is specified is very important. Another form of READ statement is

```
30 READ A$,B$
```

where $A\$$ and $B\$$ are string variables. Simple, subscripted and string variables may be combined in the same READ statement. Thus the statement

```
40 READ X,Y,A$,B$,Z(1)
```

is quite valid.

The computer is made to output data by the use of the PRINT statement. This can be used to print the value of a variable or text depending on whether or not the part of the statement after PRINT is enclosed in inverted commas. For example, the statement

```
20 PRINT A,B
```

would cause the numeric values of variables A and B to be printed out separated by a space. The alternative version

```
20 PRINT "VALUES OF A AND B = "
```

would cause the computer to print out

```
VALUES OF A AND B =
```

These two types of PRINT statements may be combined in the following form:

```
40 PRINT " VALUE OF A = ";A
```

If the variable A has previously been assigned a value of 6.32 the execution of the above statement causes the computer to print out

```
VALUE OF A = 6.32
```

Arithmetic expressions may also be included in the PRINT statement, for example,

```
40 PRINT " SQUARE OF A + B IS "; (A+B)*(A+B)
```

The PRINT statement used on its own without a following expression causes the computer to skip a line. In effect a blank line is printed and this is useful for separating parts of the output to enhance clarity.

REMARK statements enable the programmer to insert into the program material which is used for clarification but which does not form part of the program. This can be very useful in complicated programs and can assist anyone reading the program to understand the operations which are being undertaken. For example, Figure 1.7 shows a series of REMARK statements which could be added to

```
10 REM CALCULATION OF CYLINDER VOLUME
35 REM CALCULATION
45 REM PRINT DATA
55 REM TEST FOR COMPLETION
```
Figure 1.7

Figure 1.4 in order to describe the various parts of the program. As the computer ignores everything in a line following the first three letters of REMARK, this makes the letters REM interchangeable with REMARK.

1.5 Expressions, functions and subroutines

Mathematical expressions and BASIC expressions are quite similar. They consist essentially of a combination of the operations of addition, subtraction, multiplication, division, and exponentiation. The operators used to accomplish these operations are shown in Figure 1.8. Knowledge of the order in which operations are carried out, i.e. the hierarchy of operations, is important to avoid confusion. Where parentheses are used the operations contained within parentheses are performed first. Because of this it is a good idea to use parentheses if there is any doubt. Otherwise arithmetic operations are undertaken in a fixed order of precedence. Exponentiation is performed before negation. These are followed by multiplication and division and lastly by addition and subtraction. Operations of equal precedence are undertaken from left to right. The confusion which might arise if the precedence of operation is not taken into account can be illustrated by the solution of a simple quadratic equation, for example,

$$AX^2 + BX + C = 0.$$

One root of this equation is given by

$$X = \frac{-B + (B^2 - 4AC)^{\frac{1}{2}}}{2A}$$

Operator	Operation	Example
+	Addition	A + B
−	Subtraction	A − B
*	Multiplication	A*B
/	Division	A/B
↑ or ∧	Exponentiation	A↑B
−	Negation	− A

Figure 1.8 Mathematical operators

In BASIC this expression must be written using parenthesis, i.e.

```
20 LET X=(- B + SQR( B * B - 4 * A * C ))/( 2.0 * A )
```

The confusion which would exist if the parentheses were omitted is obvious. Negation consists of giving a variable a negative value. For example if X has the value 2.0, the statement

```
10 Y=-X
```

would set $y = -2.0$, and the statement

```
20 Z=-X^2.0
```

would (normally) set Z equal to -4.0. (Some lesser used forms of BASIC would return the value $+4.0$).

The primary computational statement is the LET statement used in Figures 1.1 to 1.6. The right hand side of the expression following the = sign may be any numerical value or any arithmetic expression. When the variable is a string variable, the right hand side may of course be a string. The LET statement assigns the numeric value of the expression to the variable. In many forms of BASIC the LET may be omitted so that in Figure 1.1 line 200 could read

```
200 A = 3.142*R*R
```

Figure 1.9 contains a list of library functions. These are standard functions built into BASIC and can be used directly in a LET statement. For example, the statement

```
20 LET X = SIN(Y)
```

causes the sine of the angle Y, where Y is in radians, to be computed and assigned to the variable X. Most statements are self explanatory but note that the INT function truncates the decimal part of a number and assigns a value lower than the real number specified. For example, 12.9 would be truncatd to 12 and -4.1 would be assigned a value -5. The TAB function allows the programmer to specify exactly the location of material printed out using a PRINT statement. This allows fairly precise formatting of output data. The use of a comma in formatting was discussed earlier.

If a program requires the execution of a number of statements on several different occasions, it is convenient to define these statements as a subroutine, and make use of the routine as necessary throughout

Function	Operation
ABS	Determines the absolute value
ATN	Determines the arctangent
COS	Determines the cosine (angle in radians)
COT	Determines the cotangent (angle in radians)
EXP	Raises e to a power
INT	Converts to integer form
LOG	Determines the natural logarithm
SGN	Determines the sign $+1$, 0 or -1
SIN	Determines the sine (angle in radians)
SQR	Determines the square root
TAB	Formatting function
TAN	Determines the tangent

Examples
Let $Y = \sin(X)$ sets Y equal to the sine of X
Let $Y = SQR (X)$ sets Y equal to the square root of X

Figure 1.9 Typical library functions

the main program. For example if the area of a circle was to be calculated at a number of different points in a program, then the statements listed in Figure 1.1 might be used as a subroutine. In the main program, the statement

```
10 GOSUB 100
```

would transfer control to the program written in Figure 1.1 whenever line 10 is executed. Statement 400 in Figure 1.1 would, however, be rewritten as

```
400 RETURN
```

The RETURN in line 400 transfers control from the subroutine back to the line in the main program immediately following line 10, i.e. the line in which the subroutine is called.

1.6 Control statements

As indicated earlier a BASIC program is executed in the sequence determined by the line numbers unless otherwise specified. The simplest statement for altering the sequence of execution is the GO TO or GOTO statement. Examples of use are shown in Figures 1.4 and 1.5. Whenever a GOTO statement is encountered, the computer transfers control immediately to the line number given on the right hand side of the statement. The computed GOTO statement causes control to be transferred to one of a group of statements, the particular one being chosen on the basis of the integer value of an expression. For example, the statement

```
30 ON A*B GOTO 100,200,300
```

would transfer control to statement 100, 200 or 300 depending on the value of the integer part of $A \times B$. If $A \times B = 1$ control is transferred to line 100; if $A \times B = 2$ to line 200, and if $A \times B = 3$ to line 300. Some forms of BASIC use this statement to transfer control if an error is encountered. The appropriate statement then would be:

```
10 ONERR GOTO 500
```

This statement transfers control to line 500 if an error occurs within the program. This can be convenient at times when it is desired to keep the program running instead of an error signal being flagged.

The END statement is used to indicate the end of a program and to transfer control back to the user. Until the introduction of microcomputers using BASIC, it was compulsory to finish every BASIC program with an END statement. However, some BASIC languages are now available in which the END statement is not required. The STOP statement may be used to terminate execution at any point in the program. It may appear at any point in a program and stops execution at that point.

1.7 Structured programming and control structures

The logic of long programs can be difficult to follow, particularly if GOTO statements are used indiscriminately to transfer control forwards and backwards within the program. In order to develop some structure within the program good programming techniques emphasize the use of modular structures. These structures compartmentalize the program which is written as a series of interconnected but semi-independent modules each one of which performs a specific task largely on its own. This structure makes the program much easier to write and much easier to read because each module, or compartment, may be developed and studied separately. With structured programming it is important that the modules are entered at the beginning and exited at the end. Unless absolutely necessary, GOTO statements should not be used to branch out of the middle of a module.

Structured programs can be achieved largely through the use of three control structures. These are the, IF THEN ELSE structure, the FOR NEXT loop and the WHILE WEND loop.

The IF THEN ELSE structure is a selection structure which permits selection between two alternatives. It operates by testing the validity of some expression and chooses one of two alternatives depending on the result.

The statement,

```
100 IF X=5 THEN Y=10 ELSE Y=0
```

sets $Y = 10$ when X is 5 and $Y = 0$ for any other value of X. The equality condition may be replaced by any of the other conditional operators shown in Figure 1.10.

Statement numbers may be used following THEN and ELSE (the ELSE part can be omitted) to transfer control, as was done in Figure 1.4. However that type of branching could result in an unstructured program and the WHILE WEND loop is a more useful form.

In Figure 1.4 the program continues to loop around statements 20–70 so long as V is not equal to zero. An alternative structure, which does not require GOTO statements, would continue to loop WHILE V is not zero. Figure 1.11 provides an example. Here the statements 40–70 are indented to show they are a module. The module starts with the WHILE statement and finishes with WEND (WHILE END).

The use of FOR NEXT statements in setting up a loop was demonstrated in Figure 1.6. The loop is initiated by statement 80 which sets a counter, I, equal to one. The following statements, until line 350 is

Operator	Operation
=	Equal to
< >	Not equal to
<	Less than
< =	Less than or equal to
>	Greater than
> =	Greater than or equal to

Figure 1.10 Conditional operators

```
20 V=1
30 WHILE V<>0
40       INPUT D
50       INPUT L
60       LET V=3.142*D*D*L/4.0
70       PRINT V,D,L
80 WEND
100 END
```

Figure 1.11

reached, are performed with that value of I. At line 350 the value of I is increased by one and control is transferred back to line 80. The process is repeated until I has been incremented to ten as indicated in line 80. Unless otherwise specified the running variable I will always increase by one unit. However, this can be changed as necessary. The statement

```
FOR I = 1 TO 10 STEP 2
```

would increase the value of I in units of two so that the calculation would be performed five times instead of ten. Again the indented statements identify the extent of the module.

1.8 Bibliography

WORLAND, P., *Introduction to BASIC Programming, A Structured Approach*, Houghton Mifflin, Boston (1980)

MARATECK, S. L., *BASIC*, Academic Press, New York (1986)

BARTEE T. C., *BASIC Computer Programming*, Harper and Row, New York (1981)

Chapter 2

Properties of fluids

2.1 Introduction

Before discussing the various properties of a fluid, it is useful to describe why, and how, fluids differ from solids. The basic difference is that a fluid cannot resist shear stresses. When a solid experiences a shear stress it deforms in the plane of a stress. If stressed within the elastic limit, internal forces are experienced and these will return the solid to its original shape when the shear stress is removed. Fluids are different because they have no resistance to shear stresses. When shear occurs the fluid will flow and will continue to flow until the shear stress is removed. Solids, of course, reach an equilibrium stage in which the internal forces resisting movement balance the externally applied forces. No such balance is possible in a fluid and flow occurs to eliminate any externally applied forces causing internal shear.

Fluids may be liquids or gases. The forces between the molecules of a solid are fairly strong and tend to hold the molecules together regardless of the shape or position of the solid. Solids are thus structurally fairly rigid. Molecular forces in a liquid are much weaker and, if the liquid is placed is an open container, a free surface develops in a position such that internal shear is eliminated. If the container is tilted the liquid eliminates shear as it moves to a new position. Molecular forces in a gas are very much weaker still and, if the gas is placed in a closed container, it will expand until it fills the whole of the container.

These forces are related to temperature and pressure. As the temperature rises the molecules become more agitated and the forces reduce. Thus a solid may be heated to become a liquid which when heated further will become a gas.

For example with increase in temperature:

ice → water → steam → gas
solid → liquid → vapour → gas

The distinction between vapour and gas relates to the phase. A vapour is a gas whose pressure and temperature are very close to the liquid phase.

Change of pressure has the opposite effect to change of temperature and increasing pressures reverse the process described above. For example, with increasing pressure:

gas → liquid → solid

This is why air or nitrogen may be stored in liquid form at low temperatures and high pressures.

2.2 Basic properties

A number of basic properties must be defined. These are:

Density, ρ = mass/unit volume (kg/m³)
Specific weight, γ = weight/unit volume (N/m³)
Specific volume, v = volume/unit mass = $1/\rho$
Specific gravity, s = ratio of density to density of water
(sometimes called relative density)

Note: some texts define specific volume as the volume per unit weight, i.e. the reciprocal of γ.

Force and mass are related by Newton's equation, i.e.

Force = Mass × Acceleration

From this, the Newton may be defined as

$$1\,N = 1\,kg\,m/s^2 \tag{2.1}$$

There must obviously be a relationship between specific weight and density because the weight of the mass is the force caused by gravitational attraction. Thus in Newton's equation the acceleration is the acceleration of gravity (9.81 m/s²) when force refers to weight.

$$\frac{weight}{vol} = \frac{mass}{vol} \times \text{acceleration of gravity}$$

or

$$\gamma = \rho g \tag{2.2}$$

For water, $\rho = 1000\,kg/m^3$ and, on the surface of the Earth, $g = 9.81$ m/s². Thus

$$\gamma = 1000\,\frac{kg}{m^3} \times 9.81\,\frac{m}{s^2}$$

$$= 9810\,\frac{kg\,m}{s^2}/m^3$$

and from Equation (2.1)

$$\gamma = 9810\,N/m^3 \tag{2.3}$$

These relationships may be demonstrated by calculating the values of ρ, v and s for gasoline which at 60°F has a specific weight of 7074 N/m^3

Specific gravity, $s = \dfrac{\rho_{gas}}{\rho_{water}}$

But $\rho = \gamma/g$

$\therefore \quad s = \dfrac{\gamma_{gas}}{\gamma_{water}} = \dfrac{7074}{9810} = 0.72$

Specific volume, $v = \dfrac{g}{\gamma} = \dfrac{9.81}{7074} \text{m}^3/\text{kg} = 1.386 \times 10^{-3} \text{m}^3/\text{kg}$

Density, $\rho = \dfrac{\gamma}{g} = \dfrac{1}{v} = \dfrac{1}{1.386 \times 10^{-3}} = 721.1 \text{ kg/m}^3$

2.3 Fluid compressibility

In a solid, plane stress is proportional to linear strain, i.e.

$$\sigma = E\varepsilon \tag{2.4}$$

Where σ = plane stress, E = Young's modulus and ε = linear strain = extension/original length. The coefficient of proportionality is known as Young's modulus.

In a liquid, the change in pressure is inversely proportional to the volume strain and the coefficient of proportionality is known as the bulk modulus, i.e.

$$dp = -E_v \frac{dV}{V} \tag{2.5}$$

where dp = change in pressure, dV = change in volume, V = original volume and E_v = bulk modulus. Because V and dV have the same dimensions, the bulk modulus has the same dimensions as pressure, i.e. N/m^2.

The modulus is a function of temperature and pressure but variation of E_v with pressure is low and, if the bulk modulus is assumed to be constant, the differentials in Equation (2.5) can be replaced by finite differences to give

$$\Delta p = -E_v \frac{\Delta V}{V}$$

or $\quad p_2 - p_1 = -E_v \dfrac{(V_2 - V_1)}{V_1}$

$$\therefore \quad p_2 - p_1 = E_v \frac{(V_1 - V_2)}{V_1} \tag{2.6}$$

with increasing Pressure $p_2 > p_1$. Thus $(p_2 - p_1) >$ zero and from Equation (2.6) $(V_1 - V_2) >$ zero giving:

$$V_2 < V_1$$

This volume decrease is as expected but the argument given above shows the necessity for the negative sign in Equation (2.5).

2.4 Gas laws

A perfect gas obeys the gas law

$$pv = RT \tag{2.7}$$

where $p =$ absolute pressure, $v =$ specific volume, $R =$ the gas constant, $T =$ absolute temperature.

This law is approximated by gases which are far removed from the liquid phase; for example, air.

The relationship between pressure and volume may also be expressed in the form

$$pv^n = \text{constant} \tag{2.8}$$

where $n = 1$ for isothermal (constant temperature) processes and $n = C_p/C_v$ (ratios of specific heats at constant pressure and constant volume) for adiabatic (frictionless, non-heat) processes. Differentiating Equation (2.8) leads to

$$n\,p\,v^{(n-1)}\,dv + v^n\,dp = 0$$

which, with Equation (2.5), gives

$$E_v = np$$

Thus, for isothermal processes $E_v = p$ while for adiabatic processes $E_v = kp$ $(k = C_p/C_v)$.

2.5 Viscosity of liquids

At the beginning of this chapter, it was stated that a fluid cannot withstand external shear forces to the extent that no flow will occur. There is, however, some resistance. Oil, malt, molasses and other 'sticky' liquids do tend to resist motion but all these liquids will flow if left long enough. The viscosity of a fluid is a measure of its ability to withstand shear or its resistance to flow. Fluids of high viscosity have a greater resistance than those of low viscosity. Again, a comparison can be made with solids. In solids the shear stress is proportional to the shear strain and

$$\tau = G\phi \tag{2.9}$$

where G = torsional rigidity, τ = shear stress and ϕ = shear strain. In liquids, the shear stress is proportional to the rate of shear strain and

$$\tau = \mu \, du/dy \tag{2.10}$$

where τ = shear stress and du/dy = rate of shear strain. In this equation the coefficient μ is the coefficient of proportionality and is known as the dynamic viscosity.

This can be explained by considering two flat plates separated by a liquid. If a force is applied to the top plate, it will move at some speed, u, and a shear stress will be developed between the plates. It can be shown that a liquid at a boundary moves at the speed of the boundary and does not slip. This has been verified experimentally. If the distance separating the plates is so small that the rate of change of velocity may be considered uniform, then the ratio du/dy in Equation (2.10) is given by the change in velocity between the plates divided by the distance between them. The force required to move the upper plate then obviously depends on the speed, the distance and on the coefficient of viscosity. It would be expected, for example, that with the plates separated by molasses or malt higher forces would be required than if the plates were separated by water.

In many problems the density and viscosity appear as a ratio. This ratio is called the kinematic viscosity and is given the greek symbol v, i.e.

$$v = \frac{\mu}{\rho} \tag{2.11}$$

where v = kinematic viscosity.

Different substances may be classified according to the relationship between shear stress and rate of strain. An ideal fluid, for example, represents a theoretical concept of a fluid with zero viscosity. There is no internal friction and absolutely no resistance to shear. A solid, on the other hand, does not flow and the rate of shear strain is zero regardless of the shear stress applied. Plastics are somewhat intermediate and resist a certain amount of stress before beginning to flow. In newtonian fluids, the dynamic viscosity is constant regardless of the rate of deformation; i.e. μ = constant in Equation (2.10). Non-newtonian fluids are fluids in which the dynamic viscosity varies and depends on the rate of shear strain. All of these are shown in Figure 2.1.

Because the dynamic viscosity is the coefficient of proportionality between shear stress and rate of shear strain, the dimensions of dy-

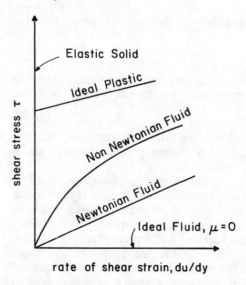

Figure 2.1 Newtonian and non-Newtonian fluids

namic viscosity are somewhat peculiar. These may be evaluated from Equation (2.10) and it can be shown that dynamic viscosity must be measured in N sec/m², 'pascal seconds'. Alternatively, the dynamic viscosity may be given in poise where:

$$1 \text{ poise} = 1 \text{ dyne s/cm}^2 \qquad (2.12)$$
$$= 10^{-1} \text{N sec/m}^2$$

The dimensions of kinematic viscosity may be similarly evaluated from Equation (2.11). These give the kinematic viscosity in terms of m²/s. These dimensions are the reverse of the dimensions of acceleration (m/s²). This provides an easy way to remember the dimensions of kinematic viscosity, and by multiplying by the dimensions of density, the dimensions of dynamic viscosity.

2.6 Surface tension

The molecular bonds in a fluid gave rise to the properties of cohesion and adhesion. These are respectively the ability to withstand tensile forces and the ability to adhere to other bodies. Together these give rise to the property known as surface tension which is the force per unit length exerted across an imaginary line drawn on the surface of the fluid.

The effects of surface tension are apparent in many different circumstances. Water spiders and other small bugs are able to walk on water because of the surface tension. Droplets from a faucet tend to form spherical shapes as they fall because of the forces developed on the surface of the drop. On smooth greasy surfaces surface tension pulls the molecules of liquid drops together to form a small globule. Small needles may be floated on top of water if they are placed in position with great care. One way of doing this is to put the needle on a sheet of toilet paper and gradually let the water soak into the toilet paper before pressing the paper under the surface. If done very carefully, the needle will remain on the surface.

Because of the ability of a liquid to adhere to a surface, the liquid will move up or down in a thin tube inserted into the fluid. Figure 2.2 shows this phenomenon which is known as capillarity. Capillary action is important in a number of situations, for example in lifting water from the root system to the upper parts of trees.

Surface tension may be evaluated experimentally by measuring capillary rise in a tube and by balancing the surface tension forces with gravitational forces. If P is the total force exerted between the fluid and the wall of the tube (see Figure 2.3), then the total force upwards is given by:

$$P \cos \theta = \pi r^2 \, \gamma \, h \tag{2.13}$$

Where θ = contact angle, r = radius of tube and h = capillary height. However,

$$P = 2\pi r \sigma$$

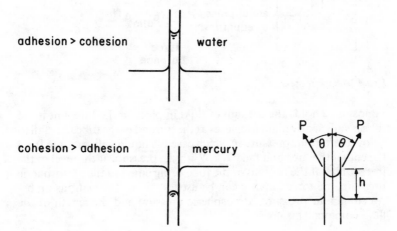

adhesion > cohesion   water

cohesion > adhesion   mercury

Figure 2.2 Effects of cohesion and adhesion   *Figure 2.3* Capillarity

and thus:

$$\sigma\, 2\pi r \cos \theta = \pi r^2 \, \gamma h$$

or $$\sigma = r\gamma h/2 \cos \theta \qquad (2.14)$$

The angle θ is approximately 0 for water in a clean glass tube.

2.7 Vapour pressure

If the pressure in a liquid drops to a sufficiently low level, the liquid will vaporize as was explained earlier. This process is analogous to boiling at low temperatures. Molecules in the vapour adjacent to the surface exert a pressure on the surface and this pressure is known as the vapour pressure. One way of visualizing this is to imagine a long tube filled with the liquid and inverted with the open end submerged in a pool of the same liquid as shown in Figure 2.4. The pressure of the

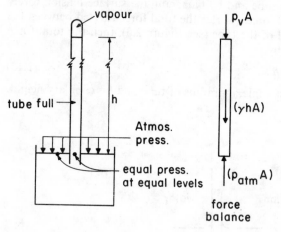

Figure 2.4 Vapour pressure

atmosphere holds the column of fluid in place in the tube but if the tube is sufficiently high, pressures at the top end of the tube may fall to below the vapour pressure in which case the liquid will separate from the end of the tube and fall back. With the pressure at the level of the free surface at the bottom of the tube being equal to the atmospheric pressure, the force balance can be used to show the relationship between vapour pressure, atmospheric pressure and the length of the fluid column. This gives:

$$p_v\, A = p_{atm}A - \gamma(hA) \qquad (2.15)$$

or

$$P_v = p_{atm} - \gamma h \qquad (2.16)$$

The device shown in Figure 2.4 is, in fact, a very simple barometer and setting $p_v = 0$ allows the atmospheric pressure to be calculated by measuring the height of the liquid column. For water, the column would be normally in excess of 10 m long. However, mercury is usually used for barometers partly because of the high value of γ leading to a mercury column less than 1 m long. Another advantage of mercury is that it has a very low vapour pressure, usually less than 0.16 pascals compared with 2500 pascals for water and over 30 000 for gasoline. This minimizes the error caused by assuming zero vapour pressure.

WORKED EXAMPLES

Example 2.1: VCYL: rotating cylinders

Two concentric cylinders are 200 mm high and have diameters of 150 mm and 156 mm. The power required to rotate the inner cylinder with the outer cylinder stationary varies with the viscosity of the fluid between them and with the speed of rotation. Write a BASIC program to obtain torque as a function of viscosity and speed. Test the program using a speed of 200 rev/min and a viscosity of 7.8×10^{-3} Pa.

```
10 REM "VCYL"
30 PRINT"THIS PROGRAM CALCULATES THE FORCE"
40 PRINT"NEEDED TO TURN A CYLINDER (R=0.075M )"
42 PRINT"WITHIN ANOTHER CYLINDER (R+0.078)"
50 PRINT
60 PRINT"ENTER SPEED OF ROTATION (REV/MIN)      ";
70 INPUT N
80 PRINT
90 PRINT" ENTER DYNAMIC VISCOSITY OF (NS/M2)      ";
100 INPUT MU
110 R1=.078
120 R2=.075
130 V1=0
140 V2=(N*2*3.1415*R2)/60
150 TAU=MU*(V2-V1)/(R1-R2)
160 F=TAU*.2*2*3.14*R2
170 TORQ=F*R2
180 POWER =2*3.142*N*TORQ/60
190 PRINT:PRINT
200 PRINT"TORQUE (NM) =    ";TORQ
210 PRINT"POWER (W)  =  ";POWER
220 PRINT:PRINT:PRINT"PROGRAM ENDED"
230 END
```

```
RUN
THIS PROGRAM CALCULATES THE FORCE
NEEDED TO TURN A CYLINDER (R=0.075M )
WITHIN ANOTHER CYLINDER (R+0.078)

ENTER SPEED OF ROTATION (REV/MIN)     ? 200

 ENTER DYNAMIC VISCOSITY OF (NS/M2)    ? 0.0078

TORQUE (NM) =     2.885312E-02
POWER (W)   =    .6043767


PROGRAM ENDED
Ok
```

Program notes

(1) Lines 10–50 print headings.
(2) Lines 60–100 input data.
(3) Lines 110–180 calculation of Equation (2.10) for shear stress.
(4) Lines 190–230 print data and end.

Example 2.2: CAPRISE: capillary rise

Develop a BASIC program to assist in plotting curves of capillary rise
for different fluids in a variety of tubes up to 1 cm diameter. As an
example use the program to plot the curve for water with a surface
tension of 0.072 N/m.

```
10 REM CAPRISE
20 PRINT"THIS PROGRAM CALCULATES THE HEIGHT OF RISE"
30 PRINT"IN CAPILLARY TUBES OF VARIOUS DIAMETERS."
40 PRINT:PRINT
50 PRINT"PLEASE ENTER THE SURFACE TENSION IN N/M   ";
60 INPUT SIGMA
70 PRINT"PLEASE INPUT THE SPECIFIC WEIGHT IN N/M3    ";
80 INPUT GAMMA
90 PRINT"PLEASE ENTER THE VALUE FOR THETA    ";
100 INPUT THETA
110 PRINT:PRINT
120 PRINT "DIAMETER (MM)         HEIGHT (MM)"
130 FOR R=.0005 TO .005 STEP .0005
140 REM step radius to .5 cm by .05 cm incr
150     H=2*SIGMA*COS(THETA)*1000/(R*GAMMA)
160        PRINT INT(R*2000);"               ";H
170 NEXT R
180 PRINT:PRINT"PROGRAM ENDED"
190 END


run
THIS PROGRAM CALCULATES THE HEIGHT OF RISE
IN CAPILLARY TUBES OF VARIOUS DIAMETERS.
```

```
PLEASE ENTER THE SURFACE TENSION IN N/M    ? 0.072
PLEASE INPUT THE SPECIFIC WEIGHT IN N/M3    ? 9810
PLEASE ENTER THE VALUE FOR THETA    ? 0

DIAMETER (MM)          HEIGHT (MM)
    1                    29.3578
    2                    14.6789
    3                     9.785932
    4                     7.33945
    5                     5.871559
    6                     4.892966
    7                     4.193971
    8                     3.669725
    9                     3.261977

PROGRAM ENDED
Ok
```

Program notes

(1) Lines 20–110 print headings and input data.
(2) Lines 120–170 calculation and print routine for nine values of diameter based on Equation (2.14).
(3) Lines 180–190 end.

Example 2.3: GCOMP: compression of a gas

A volume $0.25\,m^3$ of oxygen at 37.5°C and 17.5 kPa is compressed adiabatically to $0.01\,m^3$. What are the temperature and pressure at the end of this process? What would they have been if the compression had been isothermal? (R = 259.8 joules/kg K, $n = 1.4$)

```
10 REM "GCOMP"
30 PRINT"THIS PROGRAM CALCULATES CHANGES IN TEMPERATURE"
40 PRINT"AND PRESSURE WHEN A GAS IS COMPRESSED."
50 PRINT:PRINT
60 PRINT "PLEASE ENTER THE GAS CONSTANT     ";
70 INPUT GCONST
80 PRINT:PRINT"WHAT IS THE STARTING PRESSURE (KPa)";
90 INPUT P1
100 PRINT:PRINT"WHAT IS STARTING TEMPERATURE (C)   ";
110 INPUT T1
120 PRINT:PRINT"WHAT IS STARTING VOLUME (M3)   ";
130 INPUT VO
150 V1=GCONST*(T1+273.15)/(P1*1000)
160 PRINT:PRINT"SPECIFIC VOLUME=   ";V1
170 PRINT:PRINT"WHAT IS THE FINAL VOLUME  (M3)   ";
180 INPUT VF
190 V2=V1/(VO/VF)
200 REM CALCULATE ADIABATIC
210 P2A=P1*(V1/V2)^1.4
220 T2A=(P2A*1000*V2/GCONST)-273
230 REM CALCULATE ISOTHERMAL
240 P2I=P1*(V1/V2)
```

```
250 T2I=T1 :REM TEMPERATURE IS CONSTANT FOR ISOTHERMAL
260 REM PRINT RESULTS
270 PRINT"COMPRESSION    FINAL PRESSURE        FINAL TEMP"
280 PRINT
290 PRINT "ADIABATIC           ";P2A;"                ";T2A
300 PRINT "ISOTHERMIC          ";P2I;"                ";T2I
320 END


RUN
THIS PROGRAM CALCULATES CHANGES IN TEMPERATURE
AND PRESSURE WHEN A GAS IS COMPRESSED.

PLEASE ENTER THE GAS CONSTANT    ? 259.8

WHAT IS THE STARTING PRESSURE (KPa)? 17.5

WHAT IS STARTING TEMPERATURE (C)  ? 37.5

WHAT IS STARTING VOLUME (M3)  ? 0.025

SPECIFIC VOLUME=   4.611821

WHAT IS THE FINAL VOLUME  (M3)  ? 0.01
COMPRESSION     FINAL PRESSURE       FINAL TEMP

ADIABATIC           63.11812              175.1747
ISOTHERMIC          43.75            37.5
Ok
```

Program notes

(1) Lines 10–130 print headings and input data.
(2) Lines 150–190 specific volumes (Equation (2.7))
(3) Lines 200–220 adiabatic calculation (Equation (2.8))
(4) Lines 230–250 isothermal calculation (Equation (2.8))

PROBLEMS

(2.1) Refer to Figure 2.4 and write a program to print a table showing variation of h (for different fluids) against atmospheric pressure.
(2.2) For an atmospheric pressure of 100 kPa determine the error in h caused by assuming that the vapour pressure is zero. Consider different fluids in the analysis and print results to show the type of fluid, correct value of h and approximate value of h.
(2.3) Rewrite the program VCYL to determine the force required to pull the inner cylinder out of the outer cylinder (neither cylinder rotating) at different speeds.

Chapter 3

Fluid statics

3.1 Pressure at a point

Pressures at a point in a fluid may be represented diagramatically as shown in Figure 3.1. Notice that because the point is in a fluid, there are no shear stresses.

For equilibrium, the summation of forces in the x and in the y directions must both be xero. If the element is sufficiently small, then the weight may be considered negligible and this leads to:

$$\Sigma F_x = 0$$

$$\therefore \quad p_x \, \Delta y \, \Delta z - \left(p_N \frac{\Delta y}{\sin \theta} \Delta z \right) \sin \theta = 0$$

$$\therefore \quad p_x = p_N \tag{3.1}$$

$$\Sigma F_y = 0$$

$$\therefore \quad p_y \, \Delta x \, \Delta z - \left(p_N \frac{\Delta y}{\sin \theta} \Delta z \right) \cos \theta = 0$$

But $\quad \dfrac{\Delta_y}{\Delta_x} = \tan \theta = \dfrac{\sin \theta}{\cos \theta}$

$$\therefore \quad p_y = p_x \tag{3.2}$$

Combining Equations (3.1) and (3.2) gives:

$$p_x = p_y = p_N \tag{3.3}$$

The surface on which p_N acts was chosen arbitrarily and therefore

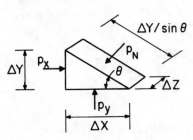

Figure 3.1 Elemental wedge

represent any plane in the fluid. This means that the pressure at a point is the same in all directions, i.e. at all values of θ.

3.2 Pressure variation with depth

Consider the pressure at some depth h below the surface of a fluid. The pressure force at this depth will be in equilibrium with the force exerted by the weight of the fluid cylinder above that depth (see Figure 3.2). Thus,

$$pA = \gamma hA \tag{3.4}$$

$$\therefore \quad p = \gamma h \tag{3.5}$$

This equation is perfectly valid when γ remains constant throughout the depth. However, to include the possibility that the specific weight or the density may vary, a somewhat more sophisticated approach is necessary. Then it is important to consider the pressure variation across a small elemental box of dimensions dx, dy and dz with the rate of variation of pressure being given by dp/dz (see Figure 3.3). Under these circumstances a force balance gives

$$-p \, dxdy + \left(p + \frac{dp}{dz} \, dz\right) dxdy + \gamma \, dxdydz = 0 \tag{3.6}$$

or

$$\frac{dp}{dz} = -\gamma \tag{3.7}$$

Note that the negative sign in this equation occurs because the positive z direction is assumed to be vertically upwards.

Both Equations ((3.5 and (3.7)) indicate that pressure increases with depth in a fluid.

The implication of these relationships may be illustrated using an example.

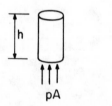

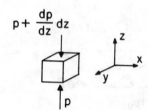

Figure 3.2 Fluid cylinder   *Figure 3.3* Pressure variation

Example

The standard atmosphere at sea level is given by:

$T = 10.8°C$, $p = 101.3$ kPa, and $\gamma = 11.93$ N/m^3

Calculate the pressure at an elevation of 3000 m assuming (a) constant density, (b) an isothermal process, and (c) an adiabatic process with $k = 1.4$.

(a) Constant density

$$\frac{dp}{dz} = -\gamma \text{ (and } \gamma = \text{constant because } \rho = \text{constant)}$$

$\therefore \quad p_2 - p_1 = -\gamma (z_2 - z_1)$

$\therefore \quad p_2 - 101.3 \times 10^3 = -11.93 \,(3000)$

$\therefore \quad p_2 = 65.5 \times 10^3$ Pa.

(b) Isothermal process

$pv = $ constant and $p/\gamma = $ constant

$\therefore \quad \dfrac{p_1}{\gamma_1} = \dfrac{p_2}{\gamma_2} = \dfrac{p}{\gamma} \quad \therefore \quad \gamma = p\dfrac{\gamma_1}{p_1}$

$$\frac{dp}{dz} = -\gamma = -p\frac{\gamma_1}{p_1}$$

Cross multiplying and integrating, with γ_1 and ρ_1 being constant:

$$\int \frac{dp}{p} = -\left(\frac{\gamma_1}{p_1}\right) \int dz$$

$$\ln(p_2/p_1) = -\frac{11.93}{101.3 \times 10^3}(3000)$$

$\therefore \quad p_2 = .7024\, p_1 = 71.15$ kPa

(c) Adiabatic process

$$pv^{1.4} = \text{constant or } \frac{p}{\gamma^{1.4}} = \text{constant}$$

$\therefore \quad$ as before

$$\frac{dp}{dz} = -\gamma = -p^{5/7} \times \left(\frac{\gamma_1}{p_1{}^{5/7}}\right)$$

$$\therefore \quad \int p^{-5/7}\, dp = -\frac{\gamma_1}{p_1{}^{5/7}} \int dz$$

$$\therefore \quad \frac{7}{2}\left(p_2^{2/7} - p_1^{2/7}\right) = \frac{11.93}{(101.3 \times 10^3)^{5/7}}(3000)$$

$$\therefore \quad p_2 = 69.9 \, \text{kPa}$$

3.3 Pressure and head

Equation (3.5) shows that the pressure at any depth h below the surface of a liquid is given by the product γh and it is therefore possible to specify pressure in terms of N/m^2 or pascals (i.e. as p) or in terms of the equivalent depth h, given by the ratio p/γ. This ratio is called the head or the pressure head. For example, at a depth of 10 m in water the pressure is given by:

$$\text{pressure } p = \gamma h$$
$$= 9810 \times 10 = 98\,100 \, \text{N/m}^2$$

alternatively

$$\text{head} = h = \rho/\gamma = 10 \, \text{m (water)} \tag{3.8}$$

In general, a pressure head of h metres of a fluid means that the pressure is the same as that experienced at a depth of h metres below the surface of the fluid. In specifying head therefore it is most important to specify the fluid in which the head is measured. In the previous example, 10 m below the surface of a water body the pressure could be specified as Newtons per square metre or metres head of water or metres head of any other fluid. To calculate the head of mercury ($s = 13.6$) the pressure remains unchanged at 98 100 N/m^2 as above but the pressure head is given by:

$$\text{head } h = \frac{p}{\gamma_{\text{merc}}} = \frac{98\,100}{13.6 \times 9810} = 0.735 \, \text{m (mercury)} \tag{3.9}$$

In terms of air (an unusual choice)

$$h = \frac{p}{\gamma_{\text{air}}} = \frac{98\,100}{11.93} = 8222 \, \text{m (air)} \tag{3.10}$$

where $\gamma_{\text{air}} = 11.93 \, \text{N/m}^2$

Pressure head and elevation are intimately related in any liquid. Consider two points in a fluid at rest as shown in Figure 3.4. From the geometry of the figure:

$$Z_1 + h_1 = Z_2 + h_2$$

But $h = \dfrac{p}{\gamma}$

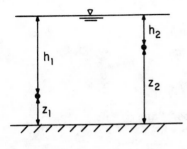

Figure 3.4 Elevation and pressure in static fluid

$$\therefore \qquad Z_1 + \frac{p_1}{\gamma} = Z_2 + \frac{p_2}{\gamma} \tag{3.11}$$

i.e. pressure head + elevation = constant in a static fluid.

Exactly the same result may be obtained from Equation (3.7) by using finite differences instead of differentials and expanding, then:

$$\Delta p = -\gamma(\Delta z) \tag{3.12}$$

or

$$p_2 - p_1 = -\gamma(z_2 - z_1) \tag{3.13}$$

3.4 Measurement of pressure

Most pressures are measured by gauges which give zero reading at atmospheric pressures. It is necessary however, to distinguish between absolute pressure, p_{abs}, atmospheric pressure, p_{atm}, and gauge pressure, p_g. Absolute pressure is the pressure measured relative to some absolute zero which would be taken to be the pressure of a true vacuum. Atmospheric pressure is the pressure exerted by the weight of the atmosphere of the Earth. This is given as an absolute pressure, i.e. relative to zero, and is normally a little more than 100 kPa (14.7 lb. per square inch, 10.3 m head of water or approximately 0.76 m head of mercury). Gauge pressure is the pressure relative to the atmosphere and may be positive if the pressure is above atmospheric or negative if the pressure is below atmospheric. These relationships are shown diagramatically in Figure 3.5.

Atmospheric pressure may be measured by an inverted tube which is called a barometer and gives a measurement of barometric pressure. This was illustrated in the previous chapter and was shown in Figure 2.4.

Various gauges are available for measuring pressures. Many use a variety of diaphragms and bellows but the most common inexpensive device is probably the Bourdon pressure gauge. This consists essentially of a tube bent into the shape of a 'C'. When pressure is applied

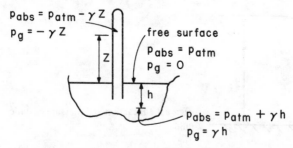

$$P_{abs} = P_{atm} - \gamma Z$$
$$P_g = -\gamma Z$$

free surface
$$P_{abs} = P_{atm}$$
$$P_g = 0$$

$$P_{abs} = P_{atm} + \gamma h$$
$$P_g = \gamma h$$

Figure 3.5 Atmospheric gauge and absolute pressures – tube full of fluid

to the inside of the tube, deformation occurs and opens the gap between the two ends of the 'C'. The movement is recorded with a fairly simple mechanical linkage.

3.5 Manometers

Manometers use the relationship between pressure and the height of a fluid column (as given in Equation (3.5.)) for measuring pressure or pressure difference. The simplest type of manometer is a piezometric tube. This is simply an open-ended tube connected to a pipe carrying fluid under pressure as shown in Figure 3.6. With the pipe under

press = p $h = p/\gamma$

Figure 3.6 Simple manometer or piezometric tube

pressure, fluid will rise up the tube until the head of fluid balances the pressure. Certain problems are however associated with the simple manometer. A long length of tube is needed for moderate pressures. Air will leak into the pipe if the pressure is less than atmospheric (i.e. gauge pressure negative) and the manometer will be of no use. Gas pressures cannot be measured using a piezometric tube because there is no free surface.

All of these difficulties can be overcome by using a U-tube containing a dense liquid which does not mix with the fluid whose pressure is being measured. Such a system is shown in Figure 3.7. A considerable variety of U-tube arrangements are possible and it is particularly important to understand the method of analysis rather than attempt to memorize formulae. The basic method consists of balancing pressures in the left-hand and right-hand legs of the U-tube at the same level in the manometer fluid. For example, for the situation shown in Figure 3.7:

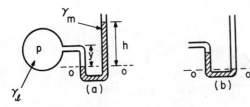

Figure 3.7 Simple U-tube

(a)	(b)
$p_{abs} + \gamma_l y = p_{atm} + \gamma_m h$	$p_{abs} + \gamma_l y + \gamma_m h = p_{atm}$
$\therefore \quad p_{abs} = p_{atm} + \gamma_m h - \gamma_l y$	$\therefore \quad p_{abs} = p_{atm} - \gamma_m h - \gamma_l y$
$\therefore \quad p_g = \gamma_m h - \gamma_l y$	$\therefore \quad p_g = -(\gamma_m h + \gamma_l y)$
$\dfrac{p_g}{\gamma_1} = \dfrac{\gamma_m}{\gamma_1} h - y$	$\therefore \quad \dfrac{p_g}{\gamma_1} = -\left(\dfrac{\gamma_m}{\gamma_1} h + y\right)$

where p_g = gauge pressure

Differential manometers are often used for measuring differences in pressure; for example, across an obstruction such as an orifice in a pipeline. With the pressure falling from p_1 to p_2 across the orifice, as shown in Figure 3.8, it is possible again to balance the pressures at the elevation of the lower surface of the manometer fluid. Then

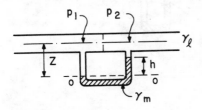

Figure 3.8 Differential U-tube

$$p_1 + \gamma_1 z = p_2 + \gamma_1(z - h) + \gamma_{mh}$$

$$\therefore \quad \frac{p_1 - p_2}{\gamma_1} = \frac{\gamma_m}{\gamma_1} h - h$$

$$= h(S - 1)$$

where S = specific gravity of liquid in pipe

Notice that the elevation of the manometer (i.e. z) relative to the pipe is not important.

For small pressure differences the U-tube may be inverted and filled with a manometer liquid less dense than the fluid in the pipe. Then from Figure 3.9

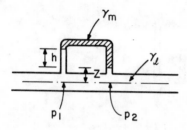

Figure 3.9 Inverted differential manometer

$$p_1 - \gamma_1(z+h) = p_2 - z\gamma_1 - h\gamma_m$$

$$\frac{p_1 - p_2}{\gamma_1} = h(1-s)$$

Some of these relationships can be demonstrated from the arrangement shown in Figure 3.10. Pressures p_1 and p_2 are measured by a

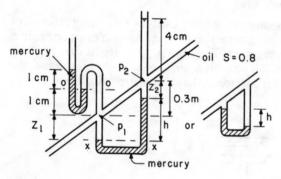

Figure 3.10 Example

U-tube manometer and a piezometric tube respectively and the difference between them is measured using a differential U-tube. If the pipe transmits light oil of a specific gravity 0.8 and the U-tubes contain mercury, specific gravity 13.6, determine the pressure difference and the differential height h. In calculating the pressure difference, express it in terms of N/mm^2 and also as equivalent heads of mercury, water and oil.

Piezometer

$$p_2 = \gamma_o h$$

$$\therefore \quad \frac{p_2}{\gamma_o} = 4 \times 10^{-2}\,m$$

U-Tube

Equating pressures at *oo*

$$p_{oo} = \gamma_m \times 10^{-2} \text{ N/m}^2$$

$$\therefore \quad p_1 = \gamma_m \times 10^{-2} + \gamma_o \times 10^{-2} \text{N/m}^2$$

and $\dfrac{p_1}{\gamma_o} = \dfrac{\gamma_m}{\gamma_o} \times 10^{-2} + 1 \times 10^{-2} = 18 \times 10^{-2} \text{ m}$

$$\therefore \quad \frac{p_1 - p_2}{\gamma_o} = 14 \times 10^{-2} \text{ m (oil)}$$

$$p_1 - p_2 = 14 \times 10^{-2} \times 9810 \times 0.8 = 1098 \text{N/m}^2 = 1.1 \times 10^{-3} \text{ N/mm}^2$$

$$\therefore \quad \frac{p_1 - p_2}{\gamma_m} = \frac{1098}{9810 \times 13.6} = 8.2 \text{ mm (mercury)}$$

and $\dfrac{p_1 - p_2}{\gamma_w} = 1098/9810 = 112 \text{ mm (water)}$

Differential U-tube

Equating pressures at *xx*:

$$p_1 + Z_1 \gamma_o = p_2 + Z_2 \gamma_o + h\gamma_m$$

$$\therefore \quad \frac{p_1 - p_2}{\gamma_o} = Z_2 - Z_1 + h\frac{\gamma_m}{\gamma_o} = 14 \times 10^{-2} \text{ m}$$

But $Z_1 + 0.3 = Z_2 + h$

$$\therefore \quad 14 \times 10^{-2} = 0.3 - h + h\left(\frac{13.6}{0.8}\right)$$

$$\therefore \quad h = -10 \text{ mm}$$

Note that the negative sign means that the assumption regarding location of manometer levels was incorrect. The correct location of the manometer fluid is shown inset in Figure 3.10.

3.6 Static forces

The force on a flat plate submerged in a liquid may be determined by considering the differential forces on a small elementary strip of the plate. This gives the force as

$$F = \gamma h_c A \tag{3.14}$$

where A = area of plate, and h_c = vertical depth from surface to centroid of plate. The centre of pressure, i.e. the point at which the force may be considered to act, is given by

$$y_p = y_c + \frac{I_c}{y_c A} \tag{3.15}$$

where y_p = distance from surface to centre of pressure and y_c = distance from surface to centroid of plate, both measured parallel to the plate. I_c = first moment of area about a centroidal axis drawn across the plate parallel to the surface.

This method is suitable only for flat plates as was described in *Basic Hydraulics* by P. D. Smith. An alternative method, equally valid for flat plates but also capable of dealing with curved and composite surfaces, is to consider the block of water adjacent to the plate as being in equilibrium and to draw a free body diagram of that block. This approach is shown in Figure 3.11 where the problem might be to calculate the force F required to keep the hinged gate closed under a head of water as shown.

The first step in this problem must be to draw the free body diagram which is shown on the right in Figure 3.11. This illustrates the block of water immediately adjacent to and above the gate. The force of the water on the gate is exerted by the weight of water in the free body and by the horizontal pressure force of the water to the left of the free body acting on the vertical surface dividing the free body from the rest of the water. These are balanced by the horizontal and vertical components at the hinge and by the force F holding the gate closed.

The weight is given by the product of specific weight and volume while the horizontal pressure force may be calculated knowing that the pressure varies from zero at the surface to γh at the bottom.

Thus the horizontal pressure force per unit width (i.e. perpendicular to the plane of the paper) is given by:

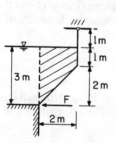
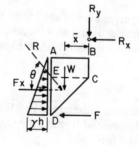

Figure 3.11 Forces on gate

$$F_x = \frac{1}{2}(0 + 3 \times 9810) \times (3 \times 1) = 44\,145 \text{ N}$$

This must act through the centroid of the pressure triangle; i.e. at a depth of $2h/3 = 2.0$ m.

Similarly, the weight, W, is given by

$$W = 9810 \times 1 \times [1 \times 2 + \frac{1}{2} \times 2 \times 2] = 32\,240 \text{ N}$$

and this acts through the centroid of the volume $ABCD$. The distance $\bar{x}$ may be obtained by area moments; i.e.

$$\text{Area } ABCD \times \bar{x} = \text{area } ABCE \times \frac{AB}{2} + \text{Area } ECD \times \frac{2}{3}EC$$

Thus $\bar{x} = 1.17$ m

The total force on the gate is the resultant of these two forces and its line of action is specified uniquely by the angle it makes with the horizontal, $\theta = \tan^{-1}(W/F_x)$. This resultant is the total force on the gate. However, for the purpose of this question it is unnecessary to calculate the total force. The horizontal restraining force at the foot of the gate may be obtained by taking moments about the hinge. Thus:

$$4F = 44\,145 \times 2 + 32\,240 \times 1.7$$
$$\therefore \quad F = 31\,502.7 \text{ N}$$

If the fluid lies under the gate as shown in Figure 3.12(a), the free body approach does not work directly but the problem may be solved by imagining fluid on both sides of the gate as if (a) and (b) were fitted together. With water on both sides of the gates to an equal level, there will be no net pressure force on the gate and so if these two diagrams are separated as shown in Figure 3.12, the forces developed in (a) must be equal and oppsoite to the forces developed in (b) which is the normal situation and which can be solved quite easily.

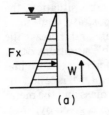
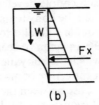

Figure 3.12 Force under gate

3.7 Buoyancy and floatation

Archimedes stated that 'When a body is immersed in a fluid it experiences an upthrust equal to the weight of the fluid displaced'. This can be shown to be true by considering the pressure variation on a small cylindrical body submerged in a liquid as shown in Figure 3.13. The vertical forces on the body vary with depth causing different pressure forces on the top and on the bottom. These are:

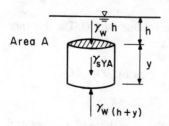

Figure 3.13 Archimedes' principle

pressure force on top $= (\gamma_w h)A$ down
pressure force on bottom $= \gamma_w (h+y)A$ up
$\therefore$ Resultant force $= \gamma_w yA =$ weight of fluid displaced.

The submerged weight of the body is given by the difference between the resultant pressure force (the upthrust) and the true weight, i.e.

submerged weight $= \gamma_s yA - \gamma_w yA$
$$= yA (\gamma_s - \gamma_w)$$

Obviously, if the specific weight of the fluid is equal to, or greater than, the specific weight of the solid then the upthrust balances or exceeds the downwards force exerted by the true weight. Under these circumstances, the body floats and displaces its own weight of the fluid in which it is floating.

The stability of a floating body is determined by the relative positions of the centre of buoyancy and the centre of mass as shown in Figure 3.14. The weight and buoyancy forces act vertically so that when the body rotates a couple develops. In a stable situation, the couple will tend to restore the body to its original position, while in an unstable situation, a couple will tend to exaggerate and increase the rotation. From Figure 3.14 it can be seen that stable situations occur when the point M (at which the buoyancy force intercepts the original vertical axis) lies above the centre of mass. This point, M, is called the metacentre and the distance between the centre of mass and the metacentre is called the metacentric height. Its location is given by:

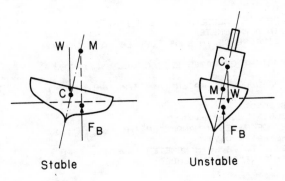

Figure 3.14 Stability

$$CM = \frac{I}{V} - CB \tag{3.16}$$

where I = first moment of area of the waterline surface about a longitudinal axis through its centroid, V = submerged volume (at rest) and B = centroid of volume V.

For stability the metacentric height must be positive, i.e. C must be below M.

WORKED EXAMPLES

Example 3.1: MANO: manometer calculations

Figure 3.15 shows a U-tube manometer connected between a closed, pressurized, water tank and an open tank containing oil of specific gravity 0.80. The specific gravity of the manometer fluid is 1.6 and the pressure in the water tank is -250 mm mercury. Determine the manometer readings and the pressure required to bring the reading to zero.

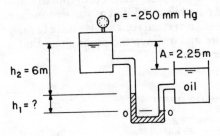

Figure 3.15 Manometer calculations

```
10 REM "MANO"
30 PRINT"THIS PROGRAM CALCULATES THE HEIGHT OF"
40 PRINT"FLUID IN A MANOMETER"
50 PRINT"ENTER DIFFERENCE IN SURF. LEVELS 'A'  (m) ";
60 INPUT A
70 PRINT"ENTER HEIGHT OF WATER SURFACE 'H2' (m)   ";
80 INPUT H2
90 PRINT"INPUT PRESSURE IN WATER TANK 'P' (mmHg)   ";
100 INPUT PRESS
110 PRESS=PRESS/1000
120 H1=(-17*PRESS)-(.25*H2)-(A)
140 PRINT"HEIGHT OF RISE IN MANOMETER 'H1'(m)  =";H1
150 PL=-((.25*H2)+A)/(.017)
160 PRINT"PRESSURE TO LEVEL MANOMETER (mmHG)  =";PL
170 PRINT:PRINT:PRINT"PROGRAM ENDED"
180 END
```

```
RUN
THIS PROGRAM CALCULATES THE HEIGHT OF
FLUID IN A MANOMETER
ENTER DIFFERENCE IN SURF. LEVELS 'A'  (m) ? 2.25
ENTER HEIGHT OF WATER SURFACE 'H2' (m)   ? 6.0
INPUT PRESSURE IN WATER TANK 'P' (mmHg)   ? -250.0
HEIGHT OF RISE IN MANOMETER 'H1'(m)  = .5
PRESSURE TO LEVEL MANOMETER (mmHG)  =-220.5882

PROGRAM ENDED
```

Program notes

(1) Lines 10–100 clear screen, print headings and input data.
(2) Lines 110–140 calculation and output based on balancing pressures at 00.
(3) Lines 150–180 calculation and output of pressure with $h=0$ in line 120.

Example 3.2: PRESS: centre of pressure

The difference in location between the centre of pressure and the centroid of a flat plate varies with depth. Develop a program to show this variaton and allow for the possibility of different ambient surface pressures. Test the program using a vertical rectangular plate (2 m by 1 m) with zero surface pressure and arrange the program so that a number of calculations can be done.

```
10 REM PRESS
20 PRINT"THIS PROGRAM GIVES THE DIFFERENCE"
30 PRINT"IN LOCATION BETWEEN THE CENTROID"
35 PRINT"AND THE CENTRE OF PRESSURE"
40 PRINT"UNDER A GIVEN DEPTH OF WATER"
```

```
50 PRINT:PRINT
60 INPUT" LENGTH OF PLATE (M) ";D
70 INPUT"ENTER WIDTH OF PLATE (M) ";B
80 INPUT"ENTER PRESS AT SURFACE (KP) ";P
90 INPUT"ENTER ANGLE OF PLATE ";THETA
100 THETA =THETA*3.142/180:P=P*1000
110 PRINT:A=0
120 WHILE A=0
130  INPUT"ENTER DEPTH TO CENTROID (M) ";Y
140  HE=(P/9180)
150  HC=(Y+HE)
160  IC=(1/12)*B*(D^3)
170  DC=(IC*(SIN(THETA)^2)/(HC*B*D))
180  PRINT"DIFF BETWEEN CENTROID AND"
190  PRINT"             CENTRE OF PRESS  = ";DC
200  INPUT"ENTER 0 TO REDO, 1 TO STOP";A
210 WEND
220 PRINT"PROGRAM ENDED"
230 END


RUN
THIS PROGRAM GIVES THE DIFFERENCE
IN LOCATION BETWEEN THE CENTROID
AND THE CENTRE OF PRESSURE
UNDER A GIVEN DEPTH OF WATER


LENGTH OF PLATE (M) ? 2.0
ENTER WIDTH OF PLATE (M) ? 1.0
ENTER PRESS AT SURFACE (KP) ? 0.0
ENTER ANGLE OF PLATE ? 90.0

ENTER DEPTH TO CENTROID (M) ? 1.0
DIFF BETWEEN CENTROID AND
             CENTRE OF PRESS  = .3333334
ENTER 0 TO REDO, 1 TO STOP? 0
ENTER DEPTH TO CENTROID (M) ? 2.0
DIFF BETWEEN CENTROID AND
             CENTRE OF PRESS  = .1666667
ENTER 0 TO REDO, 1 TO STOP? 0
ENTER DEPTH TO CENTROID (M) ? 4.0
DIFF BETWEEN CENTROID AND
             CENTRE OF PRESS  = 8.333334E-02
ENTER 0 TO REDO, 1 TO STOP? 1
PROGRAM ENDED
Ok
```

Program notes

(1) Lines 10–90 print headings and input data.
(2) Lines 100 conversion of angle to radians.
(3) Lines 110–200 calculation structure.
  Line 140 conversion of pressure to equivalent depth.
  Line 160 moment of inertia.
  Line 170 see Equation (3.15).
(4) Lines 210–230 end of program.

Example 3.3: GATEF: force on a gate

Figure 3.16 shows a circular flat gate used to separate sea water and fresh water. Determine the force necessary to open the gate for the

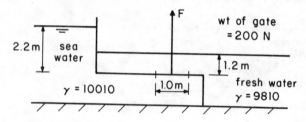

Figure 3.16 Force on a gate

conditions shown. Include in the program a routine to allow successive calculations.

```
10 REM "GATEF"
30 PRINT"THIS PROGRAM CALCULATES THE FORCE"
35 PRINT"NEEDED TO OPEN A TIDAL GATE"
40 PRINT"SEPARATING FRESH AND SALT WATER."
50 PRINT:PRINT
60 PRINT"ENTER THE WEIGHT OF THE GATE   (N)   ";
70 INPUT W
80 PRINT"ENTER THE DIAMETER OF THE GATE (M) ";
90 INPUT DIAM
100 PRINT"ENTER DEPTH OF THE FRESH WATER (M) ";
110 INPUT DF
120 PRINT"ENTER THE DEPTH OF THE SALT WATER (M) ";
130 INPUT DS
140 PRINT"ENTER SPECIFIC WT OF SALT WATER (N/M3) ";
150 INPUT GAMM1
160 PRINT"ENTER SPECIFIC WT OF FRESH WATER (N/M3) ";
170 INPUT GAMM2
180 PRINT:PRINT
190 P1=GAMM1*DS*(3.14159/4)*DIAM*DIAM
200 P2=GAMM2*DF*(3.14159/4)*DIAM*DIAM
210 F=W+P2-P1
220 PRINT"DOWNWARDS FORCE OF SEA WATER   =";P1
230 PRINT"BUOYANT FORCE OF FRESH WATER   =";P2
240 PRINT"WEIGHT OF GATE                 =";W
250 PRINT:PRINT"FORCE REQUIRED TO OPEN GATE = ";F
255 PRINT:PRINT
260 PRINT"ENTER '1' TO REDO, '0' TO END.   ";
270 INPUT A
280 IF A=1 THEN 50
290 PRINT:PRINT"PROGRAM ENDED"
300 END
```

RUN
THIS PROGRAM CALCULATES THE FORCE

```
NEEDED TO OPEN A TIDAL GATE
SEPARATING FRESH AND SALT WATER.


ENTER THE WEIGHT OF THE GATE   (N)    ? 200
ENTER THE DIAMETER OF THE GATE (M) ? 1.0
ENTER DEPTH OF THE FRESH WATER (M) ? 2.2
ENTER THE DEPTH OF THE SALT WATER (M) ? 1.2
ENTER SPECIFIC WT OF SALT WATER (N/M3) ? 10010
ENTER SPECIFIC WT OF FRESH WATER (N/M3) ? 9810


DOWNWARDS FORCE OF SEA WATER   = 9434.196
BUOYANT FORCE OF FRESH WATER   = 16950.45
WEIGHT OF GATE                 = 200

FORCE REQUIRED TO OPEN GATE =   7716.254


ENTER '1' TO REDO, '0' TO END.   ? 0
PROGRAM ENDED
Ok
```

Program notes

(1) Lines 10–180 print headings and input data.
(2) Lines 190–210 calculation of forces.
(3) Lines 220–250 output data.
(4) Lines 270–280 recalculation routine.
(5) Lines 290–300 end of program.

Example 3.4: LOCKF: force on an canal lock

A canal lock is shown in Figure 3.17. Determine the force on each gate and the reaction at the hinged supports.

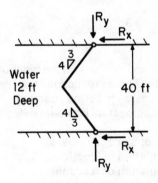

Figure 3.17

```
10 REM "LOCKF"
20 CLS:KEY OFF
30 PRINT"THIS PROGRAM CALCULATES FORCE ON A CANAL GATE."
40 PRINT" (ENTER DATA IN N,M,SEC - OR IN FT,LB,SEC.)"
50 PRINT:INPUT"ENTER DEPTH OF THE WATER       ";H
60 INPUT"ENTER WIDTH OF THE CANAL      "; W
70 INPUT"ENTER SPECIFIC WT (SI=9810, FPS=62.4)   ";GAMMA
80 INPUT"ENTER ANGLE OF GATE TO CANAL C/L";THETA
90 THETA=THETA*3.142/180
100 L=(W/2)/SIN(THETA)
110 F=GAMMA*((H^2)/2)*L
120 RX=F*SIN(THETA)
130 RY=((RX*(W/2))-((L/2)*F))/((W/2)/TAN(THETA))
140 HREACT=SQR((RX^2)+(RY^2))
150 PRINT:PRINT"LENGTH OF GATE       =";L
160 PRINT"FORCE ON GATE        =";F
170 PRINT"X COMPONENT AT GATE  =";RX
180 PRINT"Y COMPONENT AT GATE  =";RY
190 PRINT"HINGE REACTION       =";HREACT
195 IF GAMMA =62.4 THEN PRINT" ALL DATA IN FPS"
196 IF GAMMA =9810 THEN PRINT" ALL DATA IN SI"
200 PRINT:PRINT"PROGRAM ENDED"
210 END
```

```
THIS PROGRAM CALCULATES FORCE ON A CANAL GATE.
  (ENTER DATA IN N,M,SEC - OR IN FT,LB,SEC.)

ENTER DEPTH OF THE WATER      ? 12
ENTER WIDTH OF THE CANAL      ? 40
ENTER SPECIFIC WT (SI=9810, FPS=62.4)   ? 62.4
ENTER ANGLE OF GATE TO CANAL C/L? 53

LENGTH OF GATE        = 25.04045
FORCE ON GATE         = 112501.7
X COMPONENT AT GATE   = 89856
Y COMPONENT AT GATE   = 25789.12
HINGE REACTION        = 93483.58
 ALL DATA IN FPS

PROGRAM ENDED
Ok
```

Program notes

(1) Line 20 clears screen and switches off the function key descriptors at the foot of the screen. (These instructions may be incorporated in the program or typed separately before running).

(2) Lines 30–80 print headings and input data. (Compare with previous methods).

(3) Line 90 conversion of angle to radians.

(4) Lines 100–110 calculation of length and force.

(5) Lines 120–140 calculation of reactions.

(6) Lines 150–210 print data and end.

Example 3.5.: STABL: stability of a barge

A 2-m deep, 15-m long barge weighing 2×10^5 N is loaded so that the centre of gravity is at the centre line on the top surface. Determine how the stability of the barge varies with width assuming in all cases that the centre of gravity remains at the top surface. Develop a program to produce a table showing the metacentric height as a function of width. Positive values indicated stability.

```
10 REM "STABL"
20 DIM Y(10), C(10), I(10), V(10), G(10)
30 PRINT"This program will calculate the"
32 PRINT"stability of a barge with a total"
34 PRINT"weight of 2*10^5N, and a length of"
36 PRINT"15m, as it varies with ten different"
38 PRINT"values of width, read from data statements"
60 FOR I=1 TO 10
70    READ W(I)
80    Y(I)=2*10^5/(W(I)*15*9810)
90    C(I)=(2-Y(I) +(Y(I)/2))
100   I(I)=1/12*15*(W(I)^3)
110   V(I)=W(I)*15*Y(I)
120   G(I)=(I(I)/V(I))-C(I)
130   Y(I)=INT(Y(I)*100)/100
140   C(I)=INT(C(I)*100)/100
150   I(I)=INT(I(I)*100)/100
160   V(I)=INT(V(I)*100)/100
170   G(I)=INT(G(I)*100)/100
180 NEXT I
190 PRINT:PRINT:
200 PRINT "Width  Depth  Ctr B.  M I  V Disp    Gm"
210 PRINT"-----------------------------------------"
220 FOR I=1 TO 10:
230 PRINT W(I);" ";Y(I);" ";C(I);" ";I(I);" ";V(I);" ";G(I)
240 NEXT I
250 PRINT:PRINT"program ended.
260 DATA 2.1,2.5,2.6,2.9,3.1,3.3,3.6,3.8,4.1,4.7
270 END
```

```
This program will calculate the
stability of a barge with a total
weight of 2*10^5N, and a length of
15m, as it varies with ten different
values of width, read from data statements
```

```
Width   Depth   Ctr B.   M I    V Disp    Gm
-------------------------------------------------
 2.1     .64    1.67     11.57   20.38   -1.11
 2.5     .54    1.72     19.53   20.38    -.78
 2.6     .52    1.73     21.96   20.38    -.67
 2.9     .46    1.76     30.48   20.38    -.28
 3.1     .43    1.78     37.23   20.38     .04
 3.3     .41    1.79     44.92   20.38     .4
 3.6     .37    1.81     58.31   20.38    1.04
 3.8     .35    1.82     68.58   20.38    1.54
 4.1     .33    1.83     86.15   20.38    2.39
 4.7     .28    1.85    129.77   20.38    4.51

program ended.
Ok
```

Program notes

(1) Line 20 dimension statement.
(2) Lines 30–70 print headings and read widths.
(3) Lines 80–120 calculates, depth of submergence, centre of buoyancy, moment of inertia, submerged volume and location of metacentre.
(4) Lines 130–170 routine to give values to two decimal places.
(5) Lines 190–250 print output.
(6) Line 260 data.

PROBLEMS

(**3.1**) Rewrite the program MANO to determine the pressure in the water tank (in kPa and mm Mercury) for a given value of manometer reading.

(**3.2**) Rewrite the program PRESS to provide a tabular output suitable for plotting: i.e., output should be in a similar format to that given by the program STABL.

(**3.3**) In worked Example 3.4 consider the possibility that there will be salt water on one side of the lock gates and fresh water, at a different depth, on the other. Write a program to calculate the reactions.

(**3.4**) Refer to Figure 3.12 and write a program to determine the magnitude, direction and location of the force on the curved part of the gate assuming a total depth of water equal to 4.0 m. The gate is a quarter cylinder 5 m long with a radius of 1.5 m.

(**3.5**) Write a program to convert gauge pressures given in kPa, to (1) absolute pressures ($p_{atmos} = 100$ kPa), metres of water, mm of mercury and metres of an oil for which $S = 0.75$.

Chapter 4

Kinematics

4.1 Classification

Flows may be classified in a variety of ways but two useful classifications deal with consideration of the flow characteristics with respect to time and with respect to distance along the flow path. In the first of these, the flow is classified as either steady or unsteady. Steady flows are flows in which the characteristics remain constant with respect to time. Velocity, discharge and pressure etc. do not change at any particular point in the flow regardless of the time. A typical flow of this type would be the flow of water through a pipe connected between two constant head reservoirs.

Unsteady flows are those in which the characteristics of the flow change with respect to time. Almost all real flows fall into thi. category provided the flow is viewed over a sufficient length of time. The passage of a flood wave down an open channel would be a typical example. River flows vary in a yearly cycle and even flows in artificial systems such as water supply pipes vary depending on the demand for water which changes throughout the day.

Although most real flows are unsteady, it is possible to analyse many flow situations on the assumption of steady flow. This is because it is not always necessary to consider long periods of time when examining the flow. For example, in water supply systems, the pipes must be capable of dealing with the maximum flow in the system even though this may occur for a very short percentage of the time.

An alternative method of classifying the flow is to consider one instant of time and look at the variation of flow characteristics along the flow path.

In uniform flow situations, the characteristics remain constant with respect to distance along the flow. For example, the flow in a prismatic (cross sectional shape = constant) open-channel at constant depth would be considered to be a uniform flow. Relatively few natural examples of this type of flow occur but one would be the flow of water in an irrigation canal.

Varied flow is a flow in which the flow characteristics change with respect to distance along the channel. This type of flow may be sub-

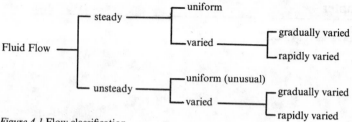

Figure 4.1 Flow classification

divided into gradually varied or rapidly varied depending on the curvature of the stream lines. Again almost all real flows are varied, with depths and velocities changing along the length of the channel or conduit.

These categories of flow are shown in Figure 4.1.

4.2 Laminar and turbulent flow

In 1883, Professor Osborne Reynolds of Manchester University discovered two distinct types of flow. He injected dye into a flowing fluid and found, when the velocity was small, that the dye travelled down the pipeline in a relatively straight unbroken line, showing that the individual particles of water moved in smooth parallel paths. Increasing the velocity caused the line to first become somewhat wavy and then to break up into numerous small eddies, the dye finally becoming diffused throughout the entire flow. This showed that the particles moved with quite irregular motion although the average motion was still in the same direction. These two types of flow have become called laminar and turbulent flow.

Laminar flow is a smooth orderly flow in which there are no eddies. The flows of viscous oil, malt or molasses would all be cases of laminar flow. These liquids are all viscous, or sticky, liquids. However, any flow may be laminar provided it occurs at a small enough velocity or at a small scale. For example, water at low velocities or around microscopic animals occurs as a laminar flow. The leak of water through a crack in a concrete tank or the flow of groundwater both tend to be viscous. The initial flow from a cigarette as the gases leave the lighted tip and move upwards in a smooth orderly path is also a laminar flow.

Turbulent flows are highly disturbed and occur with significant eddy formation. This, however, is not always readily visible to the naked eye and it is not easy to tell the difference between a turbulent flow and a laminar flow merely by looking at it. Almost all real flows are turbulent and the flow of water in rivers and canals, even though these may be perfectly clear, is turbulent. Generally, turbulent flows require either relatively high velocity, large size or low viscosity. If the gas flow from a cigarette is examined, for example, it will be seen that the laminar flow which occurs near the tip is followed by a secondary turbulent flow. This results from turbulence in the air together with the increase in size of the gas flow as it expands. These set up eddies within the flow which then ecomes turbulent.

The condition of the flow, i.e. laminar or turbulent, depends on the importance of the inertial forces relative to the viscous forces. This ratio forms a standard number called the Reynolds number which has been shown to be given by:

$$R_e = \frac{VD\rho}{\mu} = \frac{VD}{v} \propto \frac{\text{inertial forces}}{\text{viscous forces}} \qquad (4.1)$$

The foregoing discussion shows that laminar flows occur at low Reynolds numbers (when the inertial forces are small or viscous forces are large) and turbulent flows occur at large Reynolds numbers (when the inertial forces are large or viscous forces are small). If the flow in a pipe is initially turbulent and the velocity is reduced, the flow will become laminar when the Reynolds number drops to about 2000. However, if the flow is initially laminar, and the velocity is increased, the flow may remain laminar to considerably higher Reynolds numbers depending on the smoothness of the entry to the pipe and the smoothness of the pipe walls.

4.3 Continuity

Many problems in fluid mechanics may be solved by the application of three principles. These are the principles of conservation of matter, energy and momentum. Energy and momentum will be dealt with in separate chapters but the application of the principle of conservation of matter is relatively simple and may be dealt with here. Consider a pipeline such as that shown in Figure 4.2 and suppose that in time dt the fluid occupying the space BB' moves to the space CC'. Then, for matter to be conserved, the mass of fluid in the space BB' must be the same as the mass of the fluid in the space CC'. However, the fluid between the lines CB is common to both so that the mass of fluid

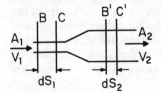

Figure 4.2 Continuity

between B and C must be the same as the mass of fluid between B' and C'. Thus,

$$\rho_1 A_1 ds_1 = \rho_2 A_2 ds_2 \tag{4.2}$$

where ρ = density and A = cross sectional area.

Dividing by dt gives

$$\rho_1 A_1 \frac{ds_1}{dt} = \rho_2 A_2 \frac{ds_2}{dt} \tag{4.3}$$

or

$$\rho_1 A_1 V_1 = \rho_2 A_2 V_2 \tag{4.4}$$

Basically this equation says that in a steady flow, the mass flow rate passing all sections is constant. If the flow is incompressible, then the density remains unchanged and can be cancelled from Equation (4.4) to give:

$$A_1 V_1 = A_2 V_2 \tag{4.5}$$

This is the equation of continuity.

Referring again to Figure 4.2, it can be seen that the line B moves to the line C over a time dt. Because B and C are separated by the distance ds_1, the volume passing B in that time is given by:

$$\text{volume} = A_1 ds_1 \tag{4.6}$$

and the rate of flow past B is given by:

$$Q = A_1 \frac{ds_1}{dt} \tag{4.7}$$

i.e. $Q = A_1 V_1 = A_2 V_2$ etc. $\tag{4.8}$

This reasoning is accurate and correct when the velocity distribution is uniform i.e. when the velocity is constant at all points across the section. In fact, this is an unreal situation and the velocity distribution is parabolic. However, Equation (4.8) is still correct provided the velocity is defined as the mean velocity across a section.

Equation (4.8) has immediate ramifications and provides the information, for example, that water does not speed up as it flows down a long constant diameter pipeline. With discharge and cross-sectional area constant, the velocity must be constant and this is true irrespective of the difference in level between the beginning and the end of the pipeline. This may be considered strange because there is undoubtedly a transfer of potential energy to some other form which is usually (in the case of a free surface or a stone rolling down a hill) considered to be kinetic energy. However, the continuity principle shows that there can be no acceleration in an enclosed pipeline of constant diameter. Instead, the potential energy is transformed into an increase in pressure as was indicated by Equation (3.11).

WORKED EXAMPLES

Example 4.1: TYPE: Type of flow

Write a program to display a table showing the roughness, k, for different types of material and to give the user the opportunity to choose a particular value of k. Then provide the necessary values of discharge etc. to calculate and print velocity, Reynolds number and relative roughness.

```
10 REM "TYPE"
20 CLS
30 PRINT "THIS PROGRAM WILL DETERMINE THE TYPE OF"
40 PRINT"FLOW BASED ON DENSITY AND MATERIAL TYPE"
50 PRINT:PRINT
60 PRINT"NO.   MATERIAL                        K (mm)"
70 PRINT
80 PRINT"1    ALUMINIUM, BRASS, PERSPEX        0.003"
90 PRINT"2    BITUMEN OR SPUN CONCRETE LINED   0.03"
100 PRINT"3    GALVANISED IRON                 0.15"
110 PRINT"4    UNCOATED CAST IRON              0.30"
120 PRINT"5    RUSTY WROUGHT IRON              0.6"
130 PRINT:INPUT"ENTER THE MATERIAL No (1-5):  ";A
140 IF A=1 THEN K=.003
150 IF A=2 THEN K=.03
160 IF A=3 THEN K=.15
170 IF A=4 THEN K=.3
180 IF A=5 THEN K=.6
190 IF K=0 THEN GOTO 20
200 INPUT"PLEASE INPUT THE DISCHARGE M3/SEC   ";Q
210 INPUT"PLEASE INPUT THE PIPE DIAMETER   M  ";D
220 INPUT"PLEASE INPUT THE DYN. VISC. NS/M2 ";MU
230 INPUT"PLEASE INPUT THE DENSITY  KG/M3    ";P
240 PRINT:PRINT:PRINT:
250 V=4*Q/(3.142*D*D)
260 REYN=(V*D*P)/MU
270 RR=K/(D*1000)
280 PRINT:PRINT"VELOCITY    =";V
290 PRINT:PRINT"REYNOLDS NUMBER   =";REYN
300 PRINT:PRINT"RELATIVE ROUGHNESS  =";RR
310 PRINT:PRINT"USE THESE VALUES WITH THE STANTON -"
320 PRINT"MOODY DIAGRAM TO DETERMINE THE TYPE "
325 PRINT"OF FLOW"
330 PRINT:PRINT:PRINT
340 PRINT"PROGRAM ENDED"
350 END
```

```
THIS PROGRAM WILL DETERMINE THE TYPE OF
FLOW BASED ON DENSITY AND MATERIAL TYPE


NO.    MATERIAL                          K (mm)

1    ALUMINIUM, BRASS, PERSPEX           0.003
2    BITUMEN OR SPUN CONCRETE LINED      0.03
3    GALVANISED IRON                     0.15
4    UNCOATED CAST IRON                  0.30
5    RUSTY WROUGHT IRON                  0.6

ENTER THE MATERIAL No (1-5):   ? 3
PLEASE INPUT THE DISCHARGE M3/SEC    ? 0.05
PLEASE INPUT THE PIPE DIAMETER  M    ? 0.4
PLEASE INPUT THE DYN. VISC. NS/M2 ? 0.001
PLEASE INPUT THE DENSITY  KG/M3      ? 1000


VELOCITY  = .3978358

REYNOLDS NUMBER    = 159134.3

RELATIVE ROUGHNESS = .000375

USE THESE VALUES WITH THE STANTON -
MOODY DIAGRAM TO DETERMINE THE TYPE
OF FLOW


PROGRAM ENDED
Ok
```

Program notes

(1) Lines 10–50 print heading.
(2) Lines 60–120 print table of k values.
(3) Lines 130–190 specification of k and error trap.
(4) Lines 200–240 input data.
(5) Lines 250–350 calculations and print output.

Example 4.2: NETW: pipe network

Figure 4.3 shows a simple pipe network with details of flow entering
and leaving. Write a program to assign flows to the various pipes

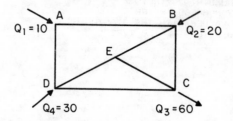

Figure 4.3 Pipe network

assuming continuity at each junction with the flow spit proportionally to the square of the diameter. Such a program might be used as a first estimate for a network analysis (see *BASIC Hydraulics*).

```
10 REM "NETW"
20 CLS:KEY OFF
30 DIM D(7)
40 PRINT"THIS PROGRAM CALCULATES FLOW THROUGH A"
50 PRINT "NETWORK OF PIPES."
60 PRINT:PRINT
70 PRINT"ENTER THE INPUT AT JUNCTION 'Q1'      ";
80 INPUT Q1
90 PRINT"ENTER THE INPUT AT JUNCTION 'Q2'      ";
100 INPUT Q2
110 PRINT"ENTER THE INPUT AT JUNCTION 'Q3'      ";
120 INPUT Q3
130 PRINT"ENTER THE INPUT AT JUNCTION 'Q4'      ";
140 INPUT Q4
150 REM read dia in order AB,AD,DE,BE,BC,DC,EC
160     FOR I= 1 TO 7
170     READ D(I)
180     NEXT I
190 QAB=Q1*(D(1)^2/((D(1)^2)+(D(4)^2)))
200 QAD=Q1-QAB
210 QDE=(Q4+QAD)*((D(3)^2)/((D(3)^2)+(D(6)^2)))
220 QDC=(Q4+QAD)-QDE
230 QBE=(Q2+QAB)*((D(4)^2)/((D(4)^2)+(D(5)^2)))
240 QBC=(Q2+QAB)-QBE
250 QEC=QDE+QBE
260 REM PRINT RESULTS
270 PRINT:PRINT
280 PRINT"PIPE        DIAMETER              FLOW"
290 PRINT"---------------------------------"
300 PRINT"AB        ";D(1)"               ";QAB
310 PRINT"AD        ";D(2)"               ";QAD
320 PRINT"DE        ";D(3)"               ";QDE
330 PRINT"BE        ";D(4)"               ";QBE
340 PRINT"EC        ";D(7)"               ";QEC
350 PRINT"DC        ";D(6)"               ";QDC
360 PRINT"BC        ";D(5)"               ";QBC
370 PRINT:PRINT
380 PRINT"PROGRAM ENDED"
390 END
400 DATA 1.0,1.5,2.0,1.0,2.0,2.5,1.75
```

```
THIS PROGRAM CALCULATES FLOW THROUGH A
NETWORK OF PIPES.


ENTER THE INPUT AT JUNCTION 'Q1'      ? 10
ENTER THE INPUT AT JUNCTION 'Q2'      ? 20
ENTER THE INPUT AT JUNCTION 'Q3'      ? 60
ENTER THE INPUT AT JUNCTION 'Q4'      ? 30
```

```
PIPE      DIAMETER        FLOW
-----------------------------------
AB        1               5
AD        1.5             5
DE        2               13.65854
BE        1               5
EC        1.75            18.65854
DC        2.5             21.34147
BC        2               20


PROGRAM ENDED
Ok
```

Program notes

(1) Lines 10–60 clear screen, keys off and print headings.
(2) Lines 7–140 input flow data.
(3) Lines 150–180 read diameter data.
(4) Lines 190–250 calculate flows using continuity.
(5) Lines 260–390 print output.
(6) Line 400 data.

Example 4.3: XSECT: cross-section area

In situations dealing with open channels it is sometimes necessary to deal with irregular shapes. These can often be simplified (see Figure 4.4) by representing them as a series of connected straight lines. The

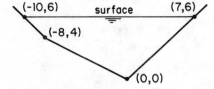

Figure 4.4 Cross-section area

shape can then be specified by giving the coordinates of the ends of the lines. In such cases 'depth' has little meaning and may be replaced in some calculations by the hydraulic mean depth (or hydraulic radius) equal to the ratio of area/wetted perimeter.

Develop a program using Figure 4.4. to determine area and perimeter for the channel shown. Use these values to calculate the hydraulic mean depth R and a Reynolds number based on $VR\rho/\mu$.

```
10 REM "XSECT"
20 ON ERROR GOTO 270
30 CLS
```

```
40 PRINT" THIS PROGRAM GIVES AREA, HYDRAULIC"
42 PRINT" MEAN DEPTH AND REYNOLDS NUMBER FOR"
44 PRINT" A X SECT DEFINED BY 4 COORDINATE POINTS"
46 PRINT"  POINT 4 IS THE POINT 0,0. NOTE THAT "
48 PRINT" Y1 AND Y3 MUST BE EQUAL"
70 PRINT" Y1 AND Y3 MUST BE EQUAL"
80 DIM X(4),Y(4) : REM  arrays for point coordinates
90 PRINT:PRINT
100 FOR I =1 TO 3
110    PRINT"For point number ";I
115    PRINT"please enter the x coordinate";
120    INPUT X(I)
130    PRINT"For point number ";I
135    PRINT"please enter the y coordinate";
140    INPUT Y(I)
150 NEXT I
160 IF Y(1)<>Y(3) THEN GOTO 70
170 A=-(Y(1)-Y(2))*.5*(X(1)+X(2))
175 A=A-.5*X(2)*Y(2)+.5*X(3)*Y(3)
180 P = SQR(((X(1)-X(2))^2)+((Y(1)-Y(2))^2))
182 P=P+SQR((X(3)^2)+(Y(1)^2))
185 P=P+SQR((X(2)^2)+(Y(2)^2))
190 R = A / P
200 PRINT:PRINT"X SECT AREA   =";A
210 PRINT:PRINT"Hydraulic Mean Depth (M)  = ";R
220 PRINT:INPUT"Please specify   discharge  m3/s ";Q
230 INPUT"Please specify the dynamic viscosity  ns/m2 ";MU
240 INPUT"Please specify   density  kg/m3  ";RHO
250 RE = (RHO*(Q/A)*R)/MU
260 PRINT:PRINT"Reynolds number  =";RE
270 PRINT:PRINT"Program Ended...."
280 END
```

```
THIS PROGRAM GIVES AREA, HYDRAULIC
MEAN DEPTH AND REYNOLDS NUMBER FOR
A X SECT DEFINED BY 4 COORDINATE POINTS
 POINT 4 IS THE POINT 0,0. NOTE THAT
Y1 AND Y3 MUST BE EQUAL


For point number  1
please enter the x coordinate? -10
For point number  1
please enter the y coordinate? 6
For point number  2
please enter the x coordinate? -8
For point number  2
please enter the y coordinate? 4
For point number  3
please enter the x coordinate? 7
For point number  3
please enter the y coordinate? 6

X SECT AREA  = 55

Hydraulic Mean Depth (M)  =  2.620015

Please specify   discharge  m3/s ? 50
```

```
Please specify the dynamic viscosity   ns/m2 ?  0.001
Please specify  density  kg/m3  ? 1000

Reynolds number  = 2381832

Program Ended....
Ok
```

Program notes

(1) Line 20 and 160 error trap to stop run.
(2) Lines 30–70 clear screen and print headings.
(3) Line 80 dimension array.
(4) Lines 100–150 specify coordinates.
(5) Lines 170–190 calculation of areas etc.
(6) Lines 200–280 print output.

PROBLEMS

(4.1.) In a river, water flows between two cross sections both having the shape shown in Figure 4.2. The first cross section, where the average velocity is 0.9 m/s, has a depth exactly as shown in Figure 4.2. Calculate the average velocity at the second cross section where the depth is half that at the first.

(4.2) Rewrite the program 'NETW' so that all data may be specified by DATA statements.

(4.3) A straight pipe 1.0 m in diameter tapers uniformly to 0.2 m diameter over a length of 30 m. The initial velocity is 0.2 m/s. Write a program to determine the discharge and the variation of velocity throughout the taper. Print results at intervals of 3.0 m along the pipe.

Chapter 5

Conservation of energy

5.1 Euler and Bernoulli equations

These equations relate to the various forms of energy which can be possessed by a fluid in motion and are an expression of the principle of conservation of energy when applied to fluids.

Consider the flow through a stream tube as shown in Figure 5.1. The stream tube is an imaginary tube within the fluid which is bound by a number of streamlines across which there is no flow. In real terms this stream tube may be considered to be a pipe.

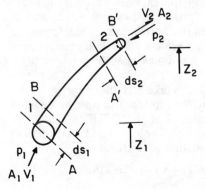

Figure 5.1 Stream tube

As work is done on the fluid the fluid moves and over a time dt the block of fluid which initially occupies the space bounded by AA′ moves to fill the space bounded by BB′. Since the space between B and A′ is common the net work done is equal to the energy of the fluid at section 2 minus the energy of fluid at section 1. However, work is given by the product of force and distance so the net work is given by

$$\text{net work} = p_1 A \, ds_1 - p_2 A_2 \, ds_2 \tag{5.1}$$

This increases the potential energy of the fluid (given by mgz) and also increase the kinetic energy (given by 0.5 mV^2). Thus,

55

$$p_1 A_1 ds_1 - p_2 A_2 ds_2 = mg(Z_2 - Z_1) + \frac{1}{2}m(V_2^2 - V_1^2) \tag{5.2}$$

But $A ds = \text{volume} = Q dt$ and $m = \text{mass} = \rho Q dt$

$$\therefore \quad p_1 Q dt - p_2 Q dt = \rho g Q dt \, (Z_2 - Z_1) + \frac{1}{2} \rho Q dt \, (V_2^2 - V_1^2)$$

and $\gamma = \rho g$

$$\therefore \quad \frac{p_1}{\gamma} - \frac{p_2}{\gamma} = (Z_2 - Z_1) + \frac{(V_2^2 - V_1^2)}{2g}$$

$$\text{or} \quad Z_1 + \frac{p_1}{\gamma} + \frac{V_1^2}{2g} = Z_2 + \frac{p_2}{\gamma} + \frac{V_2^2}{2g} \tag{5.3}$$

This equation is known as Bernoulli's equation and states merely that the energy per unit weight of fluid flowing is constant. Thus,

$$E_1 = E_2 \tag{5.4}$$

From Equation (5.3) it can be seen that the energy appears in different forms; as potential energy, Z, as kinetic energy, $V^2/2g$ and as pressure energy p/γ. It should be noted, however, that the reference to p/γ as pressure energy is really a convenience although pressure may be thought of as a form of energy in so far as it has the potential to do work.

Equation (5.4) may be expanded to take account of energy inputs (for example via a pump) energy outputs (for example by turbine) and energy losses. Variations in internal energy (INTE) might also be considered from a thermodynamics viewpoint. Then Equation (5.4) would be expanded to give an equation of the form:

$$E_1 + \text{INTE}_1 + E(\text{add}) = E_2 + \text{INTE}_2 + E(\text{out}) + E(\text{loss}) \tag{5.5}$$

Energy inputs and extractions in the form of heat might also be considered.

Equation (5.3) may also be developed by considering a force balance on an element in the stream tube. Such an approach would involve differentials and would lead to an equation of the form:

$$\frac{dp}{\rho} + V dV + g dZ = 0 \tag{5.6}$$

This equation is referred to as the Euler equation and was first derived in the middle of the eighteenth century. With the assumption that density is constant integration leads to the same result given in Equation (5.3).

5.2 Applications of Bernoulli's equation

If there are no energy losses. Equation (5.3) applies and this may be represented as shown in Figure 5.2. If small open ended vertical tubes

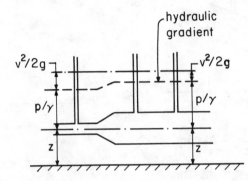

Figure 5.2 Horizontal pipe

(piezometric tubes) are connected along the pipe, the water will rise in the tubes to a level such that the weight of the column balances the pressure in the pipe. This distance (as shown in Figure 3.2) is given by p/γ. The line connecting all these levels along the pipeline is then known as the piezometric line or as the hydraulic gradient. This line provides a measure of the pressure of the fluid in the pipe. The elevation of the pipe centreline relative to some horizontal datum is Z and the distance between the hydraulic gradient and the total energy line (horizontal for no losses) must then be $V^2/2g$.

The location of the hydraulic gradient provides a measure of the gauge pressure of the fluid in the pipe and this pressure can thus always be assessed by measuring the difference in elevation between the pipe centreline and the hydraulic gradient. If the pipe is located at the hydraulic gradient, the pressure is equal to the atmospheric pressure. If the hydraulic gradient lies above the pipe, the usual case, the pressure in the pipe is positive or above atmospheric whereas if the pipe lies above the hydraulic gradient the pressure is negative or below atmospheric. Notice that in horizontal pipelines the pressure increases as the velocity decreases. Figure 5.3. shows the type of variations which might exist in a non-horizontal pipeline.

Application of Bernoulli's equation together with the continuity equation is sufficient in many cases to determine variation of velocity and pressure from one point in a pipeline to another. For example, consider the pipe shown in Figure 5.4.

For continuity

$$Q = A_1 V_1 = A_2 V_2$$

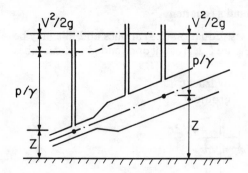

Figure 5.3 Inclined pipe

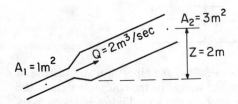

Figure 5.4 Example

$$\therefore \quad V_1 = 2/1 = 2\,\text{m/s}$$
and $V_2 = 2/3 = 0.67\,\text{m/s}$

From Bernoulli's equation

$$\frac{p_1}{\gamma} + \frac{V_1^2}{2g} + Z_1 = \frac{p_2}{\gamma} + \frac{V_2^2}{2g} + Z_2$$

$$\therefore \quad \frac{p_1}{\gamma} + \frac{4}{2 \times 9.81} + 0 = \frac{p_2}{\gamma} + \frac{(0.67)^2}{2 \times 9.81} + 2.0$$

$$\therefore \quad \frac{p_1 - p_2}{\gamma} = 1.82\,\text{m}$$

or $(p_1 - p_2) = 17.85\,\text{kPa}$

In this example, the pressure decreases by 17.85 kPa and, if either pressure is known, then the other can be calcuated. Clearly the pressure change depends partly on the variation in cross-sectional area and partly on the difference in elevation between the two points in the pipeline.

In a real flow, some energy is always lost in the direction of flow. This loss occurs primarily because of friction and turbulent eddies but minor losses also occur because of eddies set up at any abrupt changes in section and because of eddies generated by bends and valves etc. Methods of dealing with these losses are detailed in *Basic Hydraulics*.

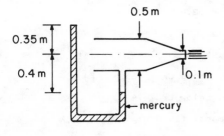

Figure 5.14 Power from a nozzle

```
5 REM "POWR"
10 CLS:KEY OFF
20 PRINT"This program calculates the power"
25 PRINT"available from a nozzle,dependent"
30 PRINT"on variations in all variables"
50 PRINT"including manometer fluid"
52 PRINT"      All data in NMS or FPS"
60 PRINT:PRINT"Enter grav. accel.( 9.81 or 32.2 )";
70 INPUT G:IF G<>9.810001 AND G<>32.2 THEN GOTO  60
80 IF G=9.810001 THEN GAMMA` =9810 ELSE GAMMA=62.4
100 PRINT:PRINT"Please enter the nozzle dia.   ";
110 INPUT D2
120 PRINT:PRINT"Please enter the pipe dia. ";
130 INPUT D1
140 PRINT:PRINT"Enter mano. surf. elevatn. above C/1";
150 INPUT H2
160 PRINT:PRINT"Enter mano. surf. elevatn. below C/L";
170 INPUT H1
180 PRINT:PRINT"Enter the spec. grav. of mano. fluid  ";
190 INPUT S
191 IF S<1! THEN PRINT "S must be GREATER than 1.0"
192 IF S<1! THEN GOTO 180
193 PRINT:PRINT
195 INPUT"      Press return for results";C$
196 CLS:PRINT:PRINT:PRINT
220 P1=(H1+H2)*S-H1
225 V2=SQR(P1*2*G/(1-((D2/D1)^4!)))
226 Q = V2*(3.1415/4)*(D2^2)
230 PO=GAMMA*Q*(P1+(V2^2)/(2*G))
235 PRINT:PRINT
240 PRINT "Results:"
245 PRINT"------------------------------------------------"
250 PRINT"Jet velocity  = ";V2
260 PRINT:PRINT"Discharge   = ";Q
270 PRINT:PRINT"Pressure head in pipe  = ";P1
280 PRINT:PRINT"Power  = ";PO
290 A$="metric"
300 IF G=32.2 THEN A$="FPS"
310 PRINT:PRINT"Results are in ";A$;" units."
320 PRINT:PRINT"Program ended.. "
330 END
```

```
This program calculates the power
available from a nozzle,dependent
on variations in all variables
including manometer fluid
        All data in NMS or FPS
```

```
Enter grav. accel.( 9.81 or 32.2 )? 9.81

Please enter the nozzle dia.   ? .1

Please enter the pipe dia. ? .5

Enter mano. surf. elevatn. above C/l? .35

Enter mano. surf. elevatn. below C/L? .4

Enter the spec. grav. of mano. fluid  ? 13.6

     Press return for results?

Results:
---------------------------------------------
Jet velocity  =  13.87747

Discharge  =  .1089902

Pressure head in pipe  =  9.800001

Power  =  20972.99

Results are in metric units.

Program ended..
Ok
```

Program notes

(1) Lines 5–80 start, print headings, enter value of g and fix γ for units chosen. Note check for correct value of g.
(2) Lines 100–193 input other data and check specific gravity.
(3) Line 195 stop screen scrolling until ready.
(4) Lines 196–230 calculations for pressure (manometer) velocity and discharge (continuity and energy).
(5) Lines 240–330 print results and stop.

Example 5.2: NOZZ: pressures in a pipe

Figure 5.15 shows a pipe connected to a reservoir. Calculate the dis-

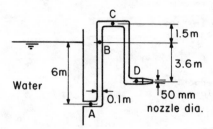

Figure 5.15 Pressures in a pipe

charge and the pressures at points A, B, C and D. Write the program
to cope with different nozzle and pipe diameters and with FPS or SI
units.

```
10 REM "NOZZ"
20 CLS:KEY OFF
30 PRINT:PRINT"This program will calculate the"
40 PRINT"flow and press at 4 points on a pipe."
50 PRINT:PRINT"Specify units (1 or 2)"
60 PRINT:PRINT"        1) g=9.81 for Metric"
70 PRINT"        2) g=32.2 for FPS"
80 INPUT A
90 IF A=1 THEN G=9.810001
100  IF A = 2 THEN G=32.2
110 IF A<>1 AND A<>2 THEN GOTO 50
120 PRINT "- The results will be in same units - "
130 PRINT:PRINT"Please enter the nozzle diameter"
140 INPUT DN
150 PRINT:PRINT"enter dist of nozz below res level"
160 INPUT ZN
170 V=SQR(ZN*2*G)
180 Q=V*3.142*DN*DN/4
190 PRINT:PRINT"Nozzle velocity = ";V
200 PRINT" Flow  = ";Q
210 PRINT:PRINT"Which point A,B,C or D  ";
220 INPUT A$
230 PRINT:PRINT"Pipe Conditions at Point ";A$
240 PRINT:PRINT"Enter pipe dia at point  ";A$;"  ";
250 INPUT DP
260 PRINT"Enter elevation at point  ";A$
270 PRINT" above res level indicate +ve, below -ve  "
280 INPUT ZP
290 VP = Q/((3.1415/4)*(DP^2))
300 HD = -ZP-(VP^2)/(2*G)
310 PRINT:PRINT"Pipe conditions at point ";A$;":"
320 PRINT:PRINT "Velocity =  ";VP
330 PRINT "Head =  ";HD
340 PRINT:PRINT"Another Calculation?  (Y/N)  ";
350 INPUT S$
360 IF S$= "y" OR S$="Y" THEN GOTO 210
370 IF S$="n" OR S$="N" THEN GOTO 390
380 GOTO 340
390 PRINT:PRINT"Program Ended..."
400 END
```

```
This program will calculate the
flow and press at 4 points on a pipe.

Specify units (1 or 2)

      1) g=9.81 for Metric
      2) g=32.2 for FPS
? 1
- The results will be in same units -

Please enter the nozzle diameter
? .05
```

```
enter dist of nozz below res level
? 3.6

Nozzle velocity =  8.404284
 Flow   =  1.650391E-02

Which point A,B,C or D  ? A

Pipe Conditions at Point A

Enter pipe dia at point  A  ? .1
Enter elevation at point  A
 above res level indicate +ve, below -ve
? -6

Pipe conditions at point A:

Velocity =   2.101406
Head =    5.774928

Another Calculation?  (Y/N)  ? N

Program Ended...
Ok
```

Program notes

(1) Lines 10–80 start, print headings and input gravitational acceleration.

(2) Lines 90–120 check value of $g = 32.2$ or 9.81 only.

(3) Lines 130–160 input other data.

(4) Lines 170–200 energy and continuity for velocity and discharge.

(5) Lines 210–280 input pipe diameter and elevation.

(6) Lines 290–330 calculate and print data.

(7) Lines 340–400 permit other calculations or stop.

Example 5.3: DUCT: Air flow in a duct

A 0.6 m square air duct bends through a radius of 1.0 m measured to the centre line of the duct. The pressure difference from the inner to the outer wall is 25 mm of water. Assuming ideal flow calculate the approximate discharge. (Take γ air $= 10.79 \, N/m^3$)

```
5 REM "DUCT"
10 CLS:KEY OFF
20 PRINT:PRINT"This program approximates the flow"
30 PRINT"bends thru a given radius."
40 PRINT:PRINT"Enter the bend radius (m)  ";
50 INPUT RAD
60 PRINT "Enter the duct width (m)  ";
70 INPUT W
80 PRINT"Enter the pressure diff head (mmH2O) ";
90 INPUT PD
```

```
95 PD=PD/1000
100 R1=RAD-(W/2)
110 R2=RAD+(W/2)
120 PR=PD*(9810/10.79)
130 C=((PR*2*9.810001)/(1/(R1^2)-(1/(R2^2))))^ .5
140 Q= W*C*LOG(R2/R1)*60
145 PRINT
150 PRINT"Flow in duct = ";Q;" cu metre per sec"
160 PRINT:PRINT"program ended..."
170 END
```

```
This program approximates the flow
bends thru a given radius.

Enter the bend radius (m)   ? 1.0
Enter the duct width (m)    ? 0.6
Enter the pressure diff head (mmH2O) ? 25

Flow in duct =   390.944  cu metre per sec

program ended...
Ok
```

Program notes

(1) Lines 5–30 start program and print headings.
(2) Lines 40–90 input data.
(3) Lines 95–140 calculations Equations (5.26)–(5.28).
(4) Lines 150–170 print data and stop.

Example 5.4: PIPE: manometer and pipe

Over a height of 1.5 m a pipe tapers uniformly from 50 mm diameter to 30 mm diameter. At a discharge Q, a mercury manometer shows a reading h. Write a program to assist development of a chart showing h as a function of Q. (The level in the manometer leg connected to the 30 mm diameter section is h m above the level in the other leg.)

```
10 REM "PIPE"
20 CLS:KEY OFF
30 PRINT:PRINT"This program will assist in developing"
40 PRINT"a chart showing the relationship between the"
50 PRINT"flow in a slanting tapered pipe and the level"
60 PRINT "of mercury in an attached manometer."
70 PRINT:PRINT
80 PRINT:PRINT"Please input the flow  (m3/s)   ";
90 INPUT Q
100 V2=4*Q/(3.142*.03*.03)
110 V1=4*Q/(3.142*.05*.05)
120 PR = 1.5 +((V2)^2 - (V1)^2/(2*9.810001))
130 H = (PR-1.5) / (13.6 - 1)
```

```
140 PRINT"The following results are for a pipe rising"
150 PRINT"1.5 metres and tapering from 50 mm to 30 mm."
160 PRINT:PRINT"Flow                          = ";Q
170 PRINT        "Downstream Velocity         = ";V2
180 PRINT        "Upstream Velocity           = ";V1
190 PRINT        "Pressure Head Difference    = ";PR
200 PRINT        "Manometer Reading           = ";H
210 PRINT:PRINT:PRINT"Program Ended...."
220 END
```

```
This program will assist in developing
a chart showing the relationship between the
flow in a slanting tapered pipe and the level
of mercury in an attached manometer.
```

```
Please input the flow  (m3/s)  ? 0.001
The following results are for a pipe rising
1.5 metres and tapering from 50 mm to 30 mm.

Flow                       =   .001
Downstream Velocity        =   1.414527
Upstream Velocity          =   .5092299
Pressure Head Difference   =   3.487671
Manometer Reading          =   .1577517


Program Ended....
Ok
```

Program notes

(1) Lines 10–90 start, print headings and input data.
(2) Lines 100–130 calculations for velocity (continuity), pressure (energy) and manometer reading.
(3) Lines 140–220 print data and stop.

Example 5.5: CAVIT: pump cavitation

Tests on a model pump indicate a critical value of the cavitation index $= 0.12$. Assuming this is constant write a program to develop a table showing variation in acceptable pump elevation relative to head.

```
10 REM "CAVIT"
20 CLS:KEY OFF
30 PRINT"    This program will develop a table"
35 PRINT"showing the variation in pump elevation"
40 PRINT"relative to the head. The critical value"
45 PRINT"of the cavitation index is"
50 PRINT"assumed to be 0.12"
55 PRINT
```

```
60 PRINT"Please enter the atmos pressure   (kPa)   ";
70 INPUT PA
75 PRINT
80 PRINT"Please enter the vapour pressure   (kPa)   ";
90 INPUT PV
100 GAMMA =9810   :REM specific weight
110 P1=(PA/GAMMA)*1000
120 P2=(PV/GAMMA)*1000
130 FOR I=1 TO 10
140 H(I)=(2+I)*10
150 Z(I)=P1-P2-.12*H(I)
160 NEXT I
170 CLS:PRINT:PRINT"Specified information:"
180 PRINT:PRINT"Specific weight = ";GAMMA
190 PRINT"Atmospheric pressure   =   ";PA
200 PRINT"Vapour Pressure    =   ";PV
210 PRINT"Cavitation parameter    =   ";.12
220 PRINT:PRINT"Head          Machine Elevation (m)"
230 PRINT"-------------------------------"
240 FOR I=1 TO 10
250 PRINT H(I),Z(I)
260 NEXT I
270 PRINT:PRINT"Program ended...."
280 END
```

    This program will develop a table
showing the variation in pump elevation
relative to the head. The critical value
of the cavitation index is
assumed to be 0.12

Please enter the atmos pressure   (kPa)   ? 101.2

Please enter the vapour pressure   (kPa)   ? 15.3

Specified information:

Specific weight =   9810
Atmospheric pressure   =   101.2
Vapour Pressure   =   15.3
Cavitation parameter   =   .12

Head	Machine Elevation (m)
30	5.156371
40	3.956371
50	2.756371
60	1.556371
70	.3563709
80	-.8436289
90	-2.043629
100	-3.24363
110	-4.44363
120	-5.643629

Program ended....
Ok

Program notes

(1) Lines 10–100 start, print headings and input data.
(2) Lines 110–160 calculation (Equation (5.21)) in loop for different heads.
(3) Lines 170–280 print table of data and stop.

PROBLEMS

(5.1) Figure 5.9 shows a pump located between two reservoirs. Write a program to determine the power of the pump based on specified values of elevation difference, flow rate, Q, diameter of pipe, D, (assumed constant over total length, L) and efficiency. Assume that the head loss, h_f, is given by the equation

$$h_f = 0.0033 \, LQ^2/D^5$$

(5.2) Write a program similar to that in Problem (1), above, but for the turbine shown in Figure 5.10.

(5.3) Write a program to determine the flow from the orifice shown in Figure 5.6. Permit variation of orifice diameter, head, coefficient of contraction and the coefficient of velocity.

(5.4) The pipe shown in Figure 5.3 has an initial diameter of 150 mm and a final diameter of 200 mm. The elevation changes from 1 m above datum to 3 m above datum and the initial pressure is 200 kPa. Calculate the final pressure in the pipe for a variety of flows ranging from 0.01 m³/s to 0.05 m³/s. Express the results in mm of Mercury.

(5.5) The flow in a horizontal rectangular channel 3-m wide is controlled by a sluice gate. The depth upstream of the gate is 2 m when the depth downstream is 0.15 m. Calculate the discharge assuming no losses through the gate.

(5.6) Develop a program based on the analysis of Problem (5) to calculate the depth downstream of the gate assuming the upstream depth and the discharge are known.

Chapter 6

Force and momentum

6.1 Force-momentum equation

When an unbalanced, or resultant, force acts on a flowing fluid, the fluid will accelerate or decelerate. This condition may be treated using Newton's second law which equates force to the product of mass and acceleration or, alternatively, in a different form, equates force to the rate of change of momentum. Momentum is the product of mass and velocity and so change in momentum may be written as:

$$\Delta M = (MV)_2 - (MV)_1 \tag{6.1}$$

The rate of change of momentum is the same as the change in momentum per unit time which, from Equation (6.1), may be written as:

$$\frac{\Delta M}{\Delta t} = \frac{MV_2}{\Delta t} - \frac{MV_1}{\Delta t} \tag{6.2}$$

But

$$\frac{M}{\Delta t} = \rho \frac{\text{vol}}{\Delta t} = \rho Q$$

and therefore:

$$F = \frac{\Delta M}{\Delta t} = (\rho Q V)_2 - (\rho Q V)_1 \tag{6.3}$$

In many cases, the discharge and density remain constant, in which case, the force can be written as:

$$F = \Delta Q \Delta V \tag{6.4}$$

where

$$\Delta V = V_2 - V_1$$

Since velocity is a vector, this equation can be applied separately in each of the possible dimensions. It should be emphasized that F is the resultant force acting on the fluid in the direction of flow and it is this force which is equal to the rate of change of momentum.

6.2 Forces on static objects

Figure 6.1 shows a 50 mm diameter jet of water being deflected through 30° while the velocity reduces to 20 m/s. The force exerted by

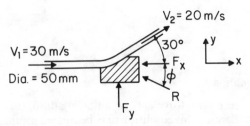

Figure 6.1 Forces on a block

the block on the fluid may be calculated by treating the problem in the x and y directions. Thus, applying Equation (6.4) in the x direction gives

$$-F_x = \rho Q (V_2 \cos \theta - V_1)$$

$$= \frac{9810}{9.81} \times \frac{\pi}{4}(.05)^2 \times 30[20 \cos 30 - 30]$$

$$\therefore \quad F_x = 747 \text{ N}$$

(Note the use of $-F_x$ because F_x acts in a direction opposite to the flow direction.)

Similarly,

$$F_y = \frac{9810}{9.81} \times \frac{\pi}{4}(0.5)^2 \times 30 (20 \sin 30 - 0) = 589 \text{ N}$$

Resultant force $= (747^2 + 589^2)^+ = 951$ N at ϕ to horizontal where $\tan \phi = \frac{589}{747}$ i.e.

$$\phi = 38.25°$$

The same sort of calculation can be applied when the jet strikes a flat plate and then splits radially. This situation is shown in Figure 6.2. When the plate is stationary, the component in the direction of the jet at the end of the momentum transfer is zero so that:

$$-F = \frac{9810}{9.81} \times \frac{\pi}{4}(.05)^2 \times 30 (0 - 30)$$

$$\therefore \quad F = 1767 \text{ N}$$

In both these calculations it is most important to adhere to the statement of the principle given in Equation (6.4), namely that the force is

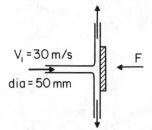

Figure 6.2 Forces on a plate

the resultant force *on* the fluid *in* the direction of flow. Notice also that the change in any quantity is given always by the second value minus the first value. In this way, it is possible to ensure the correct sign for each of the various terms and, although this is not significant in the two cases studied so far, problems can occur where the flow is enclosed and pressure forces must be taken into account. A typical example of this type would involve the calculation of the forces exerted on a pipe anchor.

When a pipeline changes direction or diameter or both, as shown in Figure 6.3, the acceleration of the fluid exerts forces on the bend which must be restricted by a concrete anchor. To determine the magnitude of these forces, it is necessary to analyse a control volume (similar to a free body diagram in statics). Here, with a number of forces acting on the volume, it is most important to ensure the correct sign in each term. Using Figure 6.3 and taking the x and y directions as indicated, the horizontal force is given by:

$$p_1 A_1 - Fx - p_2 A_2 \cos \theta = \rho Q \ (V_2 \cos \theta - V_1) \qquad (6.5)$$

Similarly, the vertical force may be written as:

$$Fy - p_2 A_2 \sin \theta = \rho Q \ (V_2 \sin \theta - 0) \qquad (6.6)$$

These equations may be solved for the unknown forces using continuity and energy equations as necessary to determine the unknown velocities and pressures.

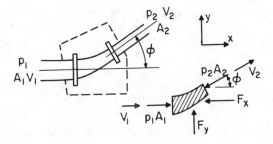

Figure 6.3 Pipe anchor

In each of the three examples given here, it has been assumed that the change of direction occurs in a horizontal plane. Were any of the three examples in a plane other than horizontal, it would be necessary to include a weight component as part of the resultant force.

6.3 Forces on moving objects

The relationships given in Equations (6.3) and (6.4) are perfectly valid for moving objects provided the discharge Q is taken to be the flow striking the object while ΔV is taken to be the change in absolute or the change in relative velocities.

Basically, what is done in these cases is to convert the unsteady flow situation to a steady situation by imagining an observer sitting on a platform and travelling parallel to and at the same speed as the object. The observer would then see a steady flow but all velocities would be the velocities relative to the object.

For example, consider again the plate shown in Figure 6.2 but with the plate moving away from the jet at 12 m/s. To an observer travelling with the plate, this situation would be identical to that shown in Figure 6.2 but with the velocity reduced from 30 m/s to $30 - 12 = 18$ m/s. The force on the plate is then given by:

$$-F = \frac{9810}{9.81} \times \frac{\pi}{4} (.05)^2 \, (18) \, [0 - 18] = 636 \text{ N}$$

In practice, the problem can be more complex when applied to the vane in a pelton turbine. Figure 6.4 shows a 50 mm diameter water jet with a velocity of 30 m/s impinging on a single vane moving in the same direction with a velocity of 18 m/s. Friction losses are such that the velocity of the water leaving the vane is 90% of that at entry.

Although it is possible to calculate the absolute velocity of the water leaving the vane using a velocity triangle, this problem is much easier to handle using the relative velocities which are shown in Figure 6.5. Then using the force momentum relationship in the x direction gives:

$$-F_x = \frac{9810}{9.81} \times \frac{\pi}{4} (.05)^2 + 12 \, [-10.8 \cos 30 - 12]$$

$$\therefore \quad F_x = 503.11 \text{ N}$$

and in the y direction:

$$F_y = \frac{9810}{9.81} \times \frac{\pi}{4} (.05)^2 \times 12 \, [10.8 \sin 30 - 0] = 127.2 \text{ N}$$

If the bucket were one of a series, for example, as would be the case

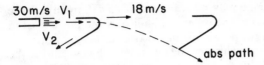

Figure 6.4 Turbine blade motion

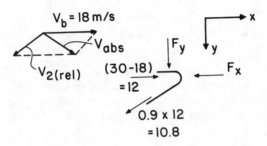

Figure 6.5 Relative and absolute velocity

in a turbine, the full flow from the jet is intercepted by the turbine because as one blade moves away, another comes in to take its place. In that case, the full flow from the 30 m/s jet must be involved in the calculation and the force in the x direction would be given by:

$$- Fx = \frac{9810}{9.81} \times \frac{\pi}{4}(.05)^2 \times 30 \ (-10.8 \cos 30 - 12)$$
$$\therefore \quad Fx = 1257.8 \text{ N}$$

Power is developed because of the forces exerted in the x direction (the y direction forces act radially) and with the buckets moving tangentially at 18 m/s, the power is given by:

Power = force × velocity
$$= 1257.8 \times 18$$
Power = 22 640 watts

As has been shown, however, the power available to the turbine is given by γQH where, in this case, H is the kinetic energy, or velocity head, of the jet impinging on the turbine, i.e. $V^2/2g$. Thus:

$$\text{Power available} = 9810 \times \left[\frac{\pi}{4} (.05)^2 \times 30 \right] \times \frac{(30)^2}{2 \times 9.81}$$
$$= 26 \ 506 \text{ watts}$$

These calculations show that the power output (22 640 watts) is, as would be expected, less than the power available (26 506 watts). The efficiency of the machine may be evaluated at 85.4% by taking the ratio of the two powers. The difference indicates the power lost in friction together with the power lost in the kinetic energy of the fluid exiting from the turbine. This may be determined by evaluating the absolute velocity of the water as it leaves the system. From the velocity triangle shown in Figure 6.5 the absolute exit velocity may be shown to be 10.2 m/s. The power lost is thus given by:

$$\text{Power lost} = 9810 \times \left[\frac{\pi}{4} (0.05)^2 \times 30 \right] \times \frac{(10.2)^2}{2 = 9.81}$$
$$= 3062 \text{ watts}$$

The difference between the power available, the power developed by the turbine and power lost at exit must represent all other losses. This loss (804 watts) would be caused by friction between the water and the vane, by friction in the bearings of the turbine and by air resistance etc.

6.4 Jet propulsion

In the enclosed tank shown in Figure 6.6, water issues as a jet with velocity V through a hole in the side of the tank. Assuming that the main body of water has zero velocity and applying the force momentum equation from the inside of the tank to the outside gives:

$$F = \rho Q (V - O) = \rho A V^2 \tag{6.7}$$

Provided the velocity is specified as the exit velocity relative to the tank, Equation (6.7) is valid whether the tank is stationary or whether the force exerted by the water on the tank causes it to move at any constant velocity. This principle is often used to propel small boats through shallow water.

The situation shown in Figure 6.6 is analogous to that of a rocket which carries all fuel within it. However, Equation (6.7) cannot be

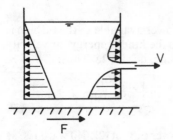

Figure 6.6 Force of jet

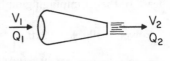

Figure 6.7 Jet engine

applied directly because the supersonic stream of gases exiting from the rocket is often at a pressure higher than atmospheric. This pressure force serves to increase the thrust of the rocket and Equation (6.7) must be transformed to:

$$F - pA = \rho A V^2 \qquad (6.8)$$

where p = jet pressure above atmospheric.

Jet engines are somewhat similar but differ in so far as they scoop air in from the atmosphere and combine this with the fuel so that the stream of exiting gases contains a mixture of fuel (which was held initially in the fuel tanks) and air (which has been added). In this situation, not only the velocity but the flow rates also change and an equation similar to Equation (6.3) must be utilized. Assuming that gases issue from the engine at atmospheric pressure, the situation is as shown in Figure 6.7 and the momentum equation leads to:

$$F = (\rho Q V)_2 - (\rho Q V)_1 \qquad (6.9)$$

But $\quad \rho Q$ = rate of mass flow = M

$\therefore \qquad F = (MV)_2 - (MV)_1$

where $\quad M_1 = M_a$ = rate of mass flow of air at entry to engine

$\qquad M_2 = M_a$ + mass of fuel consumption (M_f)

$\qquad V_1$ = air velocity at entry = velocity of aircraft

$\qquad V_2$ = velocity of exit jet

$\therefore \qquad F = (M_a + M_f) V_2 - M_a V_1$

Some comments must be made on these various relationships. First of all, each relationship is valid only for the case of steady non-accelerative motion because Newton's laws of motion cannot be applied when the reference axes are accelerating. Each equation will be altered if the exit nozzle leads to a diverging jet because in that case the particles of the jet are not moving in the same direction. Such a nozzle would be common on a rocket for example. Furthermore, the analysis given for the rocket and, to a certain extent, for the jet engine, is simplified in so far as the mass of the vehicle changes as fuel is used. The analyses given here do not take that into account.

6.5 Propellers

The thrust exerted by a propeller causes the fluid to accelerate from one steady situation to another, as shown in Figure 6.8. Application of the momentum relationship between these two states gives the force exerted by the propeller on the fluid. This is equal and opposite in direction to the pressure force exerted on the propeller by the fluid

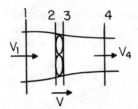

Figure 6.8 Propeller

which is given by the product of the cross-sectional area and the pressure differential. Thus:

$$F = \rho Q\,(V_4 - V_1) = (p_3 - p_2)A \tag{6.10}$$

Bernoulli's equation may be written between sections 1 and 2 and between sections 3 and 4. Bearing in mind that the pressure at sections 1 and 4 is the same, this leads to a pressure differential given by:

$$(p_3 - p_2) = \frac{1}{2}\,\rho\,(V_4^2 - V_1^2) \tag{6.11}$$

Now combining Equations (6.10) and (6.11) and writing the flow Q as the product of cross-sectional area the velocity V at the propeller:

$$V_2 = V_3 = \frac{V_1 + V_4}{2} \tag{6.12}$$

which shows that the velocity through the propeller is the arithmetic mean of the upstream and downstream velocities.

Power input is given by the sum of the power input and the wasted kinetic energy, i.e.

$$P_{in} = P_{out} + P_{kin} = \rho Q\,(V_4 - V_1)\,V_1 + \frac{\rho Q(V_4 - V_1)^2}{2} \tag{6.13}$$

Efficiency is then given by the ratio of power out to power in which, from Equation (6.13), can be shown to be:

$$\eta = \frac{P_{out}}{P_{in}} = \frac{2V_1}{(V_4 + V_1)} \tag{6.16}$$

or, from Equation (6.12),

$$\eta = \frac{V_1}{V_2} \tag{6.15}$$

WORKED EXAMPLES

Example 6.1: ANCH: force on a pipe anchor

Refer to Figure 6.3. Assuming knowledge of the geometry, discharge and upstream pressure, write a program to develop the magnitude and direction of the force on the bend.

```
10 REM "ANCH"
20 CLS:KEY OFF
30 G=9.810001
40 PRINT"THIS PROGRAM WILL CALCULATE THE FORCE ON "
45 PRINT"A HORIZONTAL PIPE ANCHOR"
50 PRINT:PRINT"G IS TAKEN AS 9.81 M/SEC/SEC"
60 PRINT:PRINT
70 INPUT"PLEASE ENTER THE FLOW (M3/S)    ";Q
80 INPUT"PLEASE ENTER DIVERGENCE ANGLE   "; THETA
90 THETA =THETA*3.14159/180
100 INPUT"ENTER THE SP WT OF FLUID (N/M3) ";GAMMA
110 INPUT"ENTER THE UPSTREAM DIAMETER  (M)  ";D1
120 INPUT"ENTER THE DOWNSTREAM DIAMETER (M) ";D2
130 INPUT"ENTER UPSTREAM PRESSURE (N/M2)    ";P1
140 REM FROM CONTINUITY
150 V1=4*Q/(3.142*D1*D1)
160 V2=V1*(D1/D2)^2
170 PRINT:PRINT"FROM CONTINUITY:"
180 PRINT"V1=    ";V1;"    M/S"
190 PRINT"V2=    ";V2;"    M/S"
200 PRINT:REM BERNOUILLI
210 P2=((P1/GAMMA)+(V1^2)/(2*G)-(V2^2)/(2*G))*GAMMA
220 PRINT"USING BERNOUILLI'S EQUATION:"
225 PRINT
230 PRINT "DOWNSTREAM PRESSURE =    ";P2;"    N/M2"
240 PRINT:REM MOMENTUM IN X DIR
250 FX=P1*3.14159*D1*D1/4-P2*3.14159*D2*D2*COS(THETA)/4
260 FX=FX+(GAMMA/G)*Q*(V1-V2*COS(THETA))
270 PRINT"USING MOMENTUM IN THE 'X' DIRECTION:"
280 PRINT"FX=    ";FX;"  (N)"
290 PRINT:REM MOMENTUM IN Y DIR
300 FY=(P2*(3.1415/4)*(D2^2)*SIN(THETA))
305 FY=FY+(GAMMA/G)*Q*(V2*SIN(THETA))
310 PRINT"USING MOMENTUM IN THE 'Y' DIRECTION:"
320 PRINT"FY=    ";FY;"  (N)"
325 PRINT
330 PRINT"RESULTANT FORCE = ";SQR((FX^2)+(FY^2));
335 PRINT"  (N)"
340 THETA2=ATN(FY/FX)*180/3.1415926#
350 PRINT"DIRECTION OF RESULTANT = ";THETA2;
355 PRINT"  DEG TO AXIS"
360 PRINT:PRINT"PROGRAM ENDED"
370 END
```

THIS PROGRAM WILL CALCULATE THE FORCE ON
A HORIZONTAL PIPE ANCHOR

G IS TAKEN AS 9.81 M/SEC/SEC

```
PLEASE ENTER THE FLOW (M3/S)      ? .5
PLEASE ENTER DIVERGENCE ANGLE   ? 30
ENTER THE SP WT OF FLUID (N/M3)  ? 9810
ENTER THE UPSTREAM DIAMETER   (M)  ? 2
ENTER THE DOWNSTREAM DIAMETER (M)  ? 1
ENTER UPSTREAM PRESSURE (N/M2)      ? 50000

FROM CONTINUITY:
V1=   .1591343    M/S
V2=   .6365373    M/S

USING BERNOUILLI'S EQUATION:

DOWNSTREAM PRESSURE =    49810.07      N/M2

USING MOMENTUM IN THE 'X' DIRECTION:
FX=    123003.9   (N)

USING MOMENTUM IN THE 'Y' DIRECTION:
FY=    19718.91   (N)

RESULTANT FORCE =  124574.5   (N)
DIRECTION OF RESULTANT =  9.107666    DEG TO AXIS

PROGRAM ENDED
Ok
```

Program notes

(1) Lines 10–60 clear screen, keys off, set value of gravitational acceleration and print headings.
(2) Lines 70–130 input data and convert angle (line 90) to radians.
(3) Lines 140–190 calculate and print velocities.
(4) Lines 200–230 calculate and print pressure.
(5) Lines 240–305 calculate force components (x and y).
(6) Lines 310–370 calculate and print resultant; stop execution.

Example 6.2.: PROP: propellor calculations

An airplane travels through still air of specific weight $12.5 \, \text{N/m}^3$ at $420 \, \text{km/hr}$. Assuming a theoretical efficiency of 88 per cent write a program to determine; (1) the air flow through the two propellers (each 2.25 m diameter), (2) the pressure difference across the propellors, (3) the thrust and (4) the power required.

```
10 REM "PROP"
20 CLS: KEY OFF
30 PRINT"  This program will calculate the power"
40 PRINT"and thrust of an aeroplane propellor"
45 PRINT
```

```
50 INPUT"Enter the propellor diameter  (m)   ";DP
55 PRINT
60 INPUT"Enter the velocity of the plane (km/h) ";VP
70 VP=VP*(10^3)/(60*60)
80 EFF=.88 :REM EFFICIENCY OF PROPELLOR IS .88
90 V=(VP/EFF)
100 Q = (V*(3.14159/4)*(DP*DP))
105 PRINT
110 PRINT "The flow through each propellor  =  ";Q
120 V4=((2*V)-VP)
130 F=((12.5/9.810001)*(2*Q)*(V4-VP))/1000
140 PD=((12.5/9.810001)*V*(V4-VP))/1000
150 PO=F*VP/EFF
160 PRINT:PRINT"Thrust  = ";F;"  kN"
170 PRINT"Pressure Difference  = ";PD;"  kPa"
180 PRINT "Theoretical Power  = ";PO;"  kW"
190 PRINT:PRINT"Program ended . . ."
200 END
```

```
    This program will calculate the power
and thrust of an aeroplane propellor

Enter the propellor diameter  (m)  ? 2.25

Enter the velocity of the plane (km/h) ? 420

The flow through each propellor  =   527.1312

Thrust  =  42.74303   kN
Pressure Difference  =  5.375027   kPa
Theoretical Power  =  5666.69   kW

Program ended . . .
Ok
```

Program notes

(1) Lines 10–60 clear screen, keys off print headings and input data.
(2) Lines 70–120 velocity and discharge calculations (equation (6.12)).
(3) Lines 130–150 calculation of thrust (6.10), pressure difference (6.11) and power.
(4) Lines 160–200 print data and stop.

Example 6.3: FOPL: force on orifice plate

Figure 6.9 shows a vertical pipe discharging water through an orifice plate. If the water between the plate and the vena contracta weighs 25 N, calculate the force on the plate.

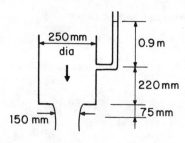

Figure 6.9 Force on orifice plate

```
10 REM "FOPL"
20 CLS: KEY OFF
30 PRINT"THIS PROGRAM WILL CALCULATE THE FORCES AND"
40 PRINT"VELOCITIES ASSOCIATED WITH AN ORIFICE PLATE"
50 PRINT
60 INPUT"ENTER THE DIAMETER OF THE PIPE   (M) ";D1
70 PRINT:PRINT" V C MEANS VENA CONTRACTA":PRINT
80 INPUT"ENTER THE DIAMETER OF THE V C  (M)  "; D2
90 PRINT:PRINT" V C MEANS VENA CONTRACTA":PRINT
100 INPUT"ENTER DIST OF PLATE FROM V C (M) "; H1
120 PRINT"ENTER DIST OF PLATE FROM PRESS TUBE (M) ";
130 INPUT H2
140 INPUT"ENTER HEAD ON PRESSURE TUBE    "; H3
160 INPUT"ENTER WT OF WATER IN JET (Kg)    "; W
180 K=(D1/D2)^2
190 V1=SQR(((H1+H2+H3)*2*9.810001)/((K^2)-1))
200 V2=K*V1
205 Q=3.142*V1*D1*D1/4
210 PRINT:PRINT"BY CONTINUITY:"
220 PRINT"   V2 = KV1, WHERE K= " ;K
230 PRINT :PRINT"BY BERNOUILLI:"
240 PRINT"UPSTREAM VELOCITY = V1 = ";V1
250 PRINT"VEL. AT VENA CONTRACTA = V2 = ";V2
255 PRINT"DISCHARGE = ";Q
260 PA=H3*9810*(3.1415926#/4)*(D1)^2
270 PRINT:PRINT"PRESSURE FORCE = PA = ";PA
280 W1=9810*(3.14159/4)*(D1^2)*H2
290 PRINT:PRINT"WEIGHT ABOVE PLATE = W1 = ";W1
300 FP=W1+W+PA-(9810/9.810001)*Q*(V2-V1)
305 PRINT:PRINT "BY MOMENTUM :"
310 PRINT"FORCE ON PLATE = ";FP
320 PRINT:PRINT:PRINT"PROGRAM ENDED"
330 END
```

```
THIS PROGRAM WILL CALCULATE THE FORCES AND
VELOCITIES ASSOCIATED WITH AN ORIFICE PLATE

ENTER THE DIAMETER OF THE PIPE   (M) ? .25

 V C MEANS VENA CONTRACTA

ENTER THE DIAMETER OF THE V C  (M)  ? .15

 V C MEANS VENA CONTRACTA
```

```
ENTER DIST OF PLATE FROM V C (M) ? .075
ENTER DIST OF PLATE FROM PRESS TUBE (M) ? .22
ENTER HEAD ON PRESSURE TUBE    ? .9
ENTER WT OF WATER IN JET (Kg)    ? 25
BY CONTINUITY:
   V2 = KV1, WHERE K=  2.777778

BY BERNOUILLI:
UPSTREAM VELOCITY = V1 =  1.868429
VEL. AT VENA CONTRACTA = V2 =  5.19008
DISCHARGE =  9.172818E-02

PRESSURE FORCE = PA =  433.3925

WEIGHT ABOVE PLATE = W1 =  105.9403

BY MOMENTUM :
FORCE ON PLATE =  259.6439


PROGRAM ENDED
Ok
```

Program notes

(1) Lines 10–60 clear screen, keys off, print headings and enter data.
(2) Lines 180–205 calculation of velocity and discharge (based on continuity and energy).
(3) Lines 210–255 print data.
(4) Lines 260–290 calcuation of pressure and weight of fluid in pipe.
(5) Lines 300–330 calculation and print force on plate; stop execution.

Example 6.4: TURP: turbine power

Referring to Section 6.3 and Figures 6.4/6.5, write a program to determine the power developed by the turbine, the power available to the turbine and the efficiency.

```
5 REM "TURP"
10 CLS : KEY OFF
20 PRINT"THIS PROGRAM WILL CALCULATE THE POWER "
25 PRINT"DEVELOPED BY A TURBINE"
30 PRINT:PRINT
40 INPUT"ENTER THE SPEED OF BUCKETS  (m/s)    ";V1
50 INPUT"ENTER THE JET VELOCITY  (m/s)    "; V2
60 INPUT"ENTER THE DIAMETER OF THE JET (m)    ";D
70 PRINT"ENTER THE ANGLE OF THE BUCKETS (deg) ";
75 INPUT THETA
80 THETA=THETA*3.14159/180
90 V3=(V2-V1)
100 PRINT:PRINT"RELATIVE ENTRY VELOCITY = ";V3
```

```
110 V4=.9*V3
120 PRINT:PRINT"RELATIVE EXIT VELOCITY = ";V4
130 F=1000*(3.1415926#/4)*D^2*V2*(V4*COS(THETA)+V3)
140 PRINT:PRINT"FORCE ON BLADE SERIES = ";F
150 PD=F*V1
160 PRINT:PRINT"POWER DEVELOPED BY TURBINE = ";PD
170 PA=9810*(3.14159/4)*D^2*V2*(V2^2)/(2*9.810001)
180 PRINT:PRINT"POWER AVAILABLE TO TURBINE = ";PA
190 EFF=PD/PA*100
200 PRINT:PRINT"EFFICIENCY = ";EFF
210 PRINT:PRINT"PROGRAM ENDED"
220 END
```

```
THIS PROGRAM WILL CALCULATE THE POWER
DEVELOPED BY A TURBINE


ENTER THE SPEED OF BUCKETS   (m/s)   ? 18
ENTER THE JET VELOCITY   (m/s)   ? 30
ENTER THE DIAMETER OF THE JET (m)   ? .05
ENTER THE ANGLE OF THE BUCKETS (deg)? 30

RELATIVE ENTRY VELOCITY =   12

RELATIVE EXIT VELOCITY =   10.8

FORCE ON BLADE SERIES =   1257.8

POWER DEVELOPED BY TURBINE =   22640.4

POWER AVAILABLE TO TURBINE =   26507.17

EFFICIENCY =   85.41238

PROGRAM ENDED
Ok
```

Programme notes

(1) Lines 5–80 clear screen, keys off, input data and convert theta to radians.
(2) Lines 90–120 velocity calculation.
(3) Lines 130–140 calculate and print force (Fx).
(4) Lines 150–160 calculate and print power.
(5) Lines 170–200 calculate and print power available and efficiency.
(6) Lines 210–220 stop execution.

PROBLEMS

(**6.1**) A 30 mm diameter jet of water moving at 50 m/s strikes a plate which splits the jet into two equal streams. Both streams are bent through 150° so that the exit flow moves in a direction opposite to that entering. Calculate the force on the plate.

(6.2) A 150 mm diameter pipe is attached to a 50 mm diameter nozzle by a flanged connection held together with four bolts. Calculate the force on the bolts when the flow is 0.1 m³/s.

(6.3) A small boat is driven by a jet propulsion system which takes water in at the bow and discharges through a nozzle at the stern. Investigate the range of thrusts which can be achieved at different boat speeds for a variety of discharge nozzles and for varying flow through the system.

(6.4) Rewrite the program ANCH for a pipe bending in the vertical plane assuming that the downstream pressure is known. Arrange the program to handle SI or FPS units.

Chapter 7

Viscous flow

7.1 Laminar flow between parallel plates

Laminar, or viscous, flow occurs when viscous forces are large relative to inertial forces. This type of flow may occur between flat parallel plates provided the plates are close together and the velocity of the fluid is small or the viscosity of the fluid is high. The analysis is fairly simple and gives approximate values for the leakage of fluid through small cracks.

The flow situation is represented by Figure 7.1 which shows fluid

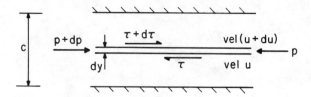

Figure 7.1 Flow between parallel plates

moving from left to right between two parallel plates of infinite extent set perpendicular to the plane of the paper. Flow occurs because of the pressure difference (dp) and the analysis may be undertaken by considering a small strip of fluid of height dy and length dx between the two plates. Variation of velocity (du) across this strip sets up shear stresses (τ) and the flow may be analyzed using a force balance. Thus:

$$(p+dp)\,dy+(\tau+d\tau)\,dx=\tau dx+pdy$$

$$\therefore \quad d\tau dx+dpdy=0$$

$$\therefore \quad \frac{d\tau}{dy}+\frac{dp}{dx}=0 \tag{7.1}$$

But

$$\tau=\mu\frac{du}{dy}$$

90

Substitution in Equation (7.1) gives:

$$\frac{dp}{dx} + \frac{d}{dy}\left(\mu\frac{du}{dy}\right) = 0$$

Integrating twice then leads to:

$$= \frac{dp}{dx}\frac{y^2}{2} + \mu u + Ay + B = 0 \tag{7.2}$$

where A and B are constants of integration.
With both plates stationary and with no slip at the boundary

$u = 0$ when $y = 0$ or $y = c$ (distance between plates)

Then

$$u = \frac{1}{2\mu}\frac{dp}{dx}(cy - y^2) \tag{7.3}$$

This indicates a parabolic profile with maximum velocity in the centre. Discharge may be evaluated from

$$Q = \int ub\, dy \tag{7.4}$$

where b is the width of the plate (perpendicular to the plane of the paper in Figure 7.1).
With the value of u given in Equation (7.3) this leads to:

$$Q = \frac{bc^3}{12\mu}\left(\frac{dp}{dx}\right)$$

$$= \frac{bc^3}{12\mu l}(p_2 - p_1) \tag{7.5}$$

where l is the length of the plate and p_1 and p_2 are the end pressures. While this is valid for horizontal plates Equation (7.5) must be modified if there are changes in elevation. This can be done simply by replacing the pressure term p by a pressure plus potential head term, i.e. $p + \gamma z$.

If the lower plate is stationary and the upper plate moves at velocity V, the boundary conditions change so that

$u = 0$ when $y = 0$ and $u = V$ when $y = c$

Substitution in Equation (7.3) then gives:

$$u = \frac{1}{2y}\left(cy - y^2\right)\frac{dp}{dx} + \frac{Vy}{c} \tag{7.6}$$

Discharge may again be obtained from Equation 7.4).

7.2 Laminar flow in circular tubes

The analysis of viscous flow in circular tubes is generally similar to that undertaken for flat plates except that the force balance must be based on a small anulus of thickness dr set at some intermediate radius r from the centre of the tube. Using the notation shown in Figure 7.2, the pressure force acting on this ring, or anulus, may be equated to the shear force to give:

$$(p + dp)\,\pi r^2 - p\pi r^2 = \tau.2\pi\,r dx \tag{7.7}$$

$$\therefore \tau = \frac{r}{2}\frac{dp}{dx} = \mu\frac{du}{dr}$$

Integrating with respect to r gives:

$$u_r = \frac{1}{2\mu}\frac{dp}{dx}\int_o^R r dr$$

$$= \frac{1}{4\mu}\frac{dp}{dx}(R^2 - r^2) \tag{7.8}$$

For no slip at the boundary $u = 0$ when $r = R$. Maximum velocity occurs in the centre where $r = 0$. Thus

$$U_{max} = \frac{1}{4\mu}\frac{dp}{dx}R^2 \tag{7.9}$$

Discharge is obtained from:

$$Q = \int_o^R u_r\,2\pi r\,dr$$

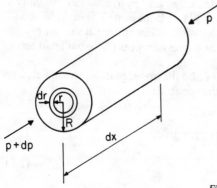

Figure 7.2 Flow through circular tube

$$= \frac{\pi R^4}{8\mu} \frac{dp}{dx}$$

$$= \frac{\pi R^4 (p_2 - p_1)}{8\mu L} \tag{7.10}$$

Here again the pressure term must be modified if changes in elevation occur.

Average velocity is given by dividing discharge by cross sectional area so that:

$$V_{av} = \frac{R^2}{8\mu} \frac{dp}{dx} \tag{7.11}$$

and if the pressure difference is caused by a head difference, h, exerted over a length L,

$$\frac{dp}{dx} = \frac{\rho g h}{L}$$

Substituting in Equation (7.11) gives

$$h = 32 \frac{\mu V L}{\rho g D^2} \tag{7.12}$$

This is known as Poiseuille's equation and provides a means of calculating the head lost due to laminar flow in a circular pipe of diameter D.

7.3 Stoke's Law

Stoke's Law refers to viscous flow around a sphere totally immersed in the fluid. It is not important whether the fluid moves past a stationary sphere or whether the sphere moves through a stationary fluid but the law is most commonly used for the situation of a small sphere falling at its terminal velocity through the fluid.

Viscous drag on a submerged body may be related to other variables using the basic definition of viscosity:

$$\tau = \mu \frac{du}{dy}$$

Shear stress is proportional to force x area and, in Chapter 9, the viscous drag, F_v, is shown to be given by

$$F_v \propto \mu v l \tag{9.35}$$

where l is some standard length.

This relationship was developed by Stokes for a rigid sphere falling through an infinite fluid in such a way that viscous forces were very large relative to inertial forces. The limit was for Reynolds numbers less than 0.10. On this basis he showed that

$$F_v = 3\pi \, \mu u d$$

where d = diameter of sphere and u = terminal velocity.

This force may be equated to the submerged weight of the sphere (i.e. total weight – buoyant upthrust). Thus:

$$3\pi \, \mu u d = \frac{\pi}{6} \, d^3 g \, (\rho_s - \rho_w) \tag{7.13}$$

where ρ = density (s of sphere, w of fluid). Solving for velocity gives

$$u = \frac{d^2 g \, (\rho_s - \rho_w)}{18 \, \mu} \tag{7.14}$$

This is Stoke's Law valid, as stated earlier, for a Reynolds number less than 0.01.

7.4 Flow in porous media

Many examples of this type of flow exist but the most common is undoubtedly the flow of ground water through soils and rock. Ground water accounts for almost two-thirds of the world's fresh water resources. Movement through the soil occurs at a very slow rate, perhaps several metres per week, and this low velocity together with the very small size of the flow passages causes the movement to occur in a viscous or liminar manner.

The problem was first studied by a French engineer, Henri Darcy, who discovered that the average velocity through a soil was proportional to the hydraulic gradient i.e.

$$Q = K \frac{dh}{dl} A \tag{7.15}$$

or $$\frac{Q}{A} = V_{av} = K \frac{dh}{dl} \tag{7.16}$$

where dh/dl = rate of head loss.

The coefficient of proportionality K is known as the hydraulic conductivity and considerable effort has been expended in attempts to develop analytical relationships describing this variable.

Experiments conducted in ideal porous media consisting of uniform closely packed spheres indicated that the conductivity was proportional to the square of the sphere diameter and to the specific weight of the fluid while there was an inverse relationship with the dynamic viscosity. Thus:

$$K = \frac{C\, d^2 \rho g}{\mu} \tag{7.17}$$

The product Cd^2 is known as the specific or intrinsic permeability and depends only on the medium through which the flow occurs, In a real, as opposed to an ideal, situation, the diameter of the grains, d, must be taken as some size which is representative of all the grains in the non-uniform media and the coefficient C must in some way depend on the shape and on the manner in which the grains are packed together. This latter quantity may be represented by the porosity, given by the ratio of the void volume to the total volume (voids + solids).

Experimental studies to determine the relationship between C and porosity have resulted in a variety of predictor equations. Of these, the best known is probably the Kozeny-Carmen equation which has the form:

$$K = \frac{\rho g}{\mu} \frac{n^3}{(1-n)^2} \left(\frac{d^{m^2}}{180} \right) \tag{7.18}$$

where n = porosity and d^m = representative grain size.

Equations (7.16) to (7.18) provide a basis for the analysis of groundwater flow in natural non-uniform soils. More information on this topic is provided in the text *BASIC Hydraulics*.

7.5 Measurement of viscosity

There are basically three different methods of measuring viscosity. One involves laminar flow in a circular tube, the second invokes Stoke's Law and the third uses the shear stress developed on a rotating cylinder.

If two reservoirs are connected by a long small diameter pipe and set at different levels to give a continuous steady flow Poiseulle's Equation (7.12) will apply provided the flow is laminar. In such a situation h is given by the difference in elevation between the reservoirs and the average velocity may be determined from measurements of discharge and cross sectional area. Equation 7.12 may then be used to calculate the viscosity of the fluid flowing in the tube. Care should be taken to check the Reynolds number to ensure laminar flow and to make the tube sufficiently long for end effects to be negligible.

Similar principles have been adopted for commercial viscometers. In some, the laminar flow occurs in a U-tube which contains a capillary section. Velocity is not measured directly. Instead the time taken for a fixed volume of fluid to fall from one mark in the tube to another is noted and viscosity is correlated against time. In the Redwood Saybolt and Engler viscometers a fixed volume of liquid is placed in a container which is itself immersed in a water bath to maintain a constant temperature. During the test, the time taken for a fixed volume to escape from the container through a capillary tube in the bottom is noted. Viscosity is then obtained from precalibrated correlation charts.

Although the principles involved in Poiseuille's equation apply to these industrial viscometers, the equation itself does not because end effects and surface tension effects may be significant. Industrial viscometers of this type must therefore be calibrated using liquids of known viscosity before being put into practical use.

The methods discussed above are not suitable for liquids of high viscosity because of the very small velocities and long flow times. Instead the viscosity of such liquids may be measured using Stoke's Law. In that method, a small sphere of known weight is allowed to fall through the fluid and the time taken to traverse a set distance is measured. Care must be taken to make measurements only after terminal velocity has been reached and for this reason, the initial mark on the cylinder is set some distance below the surface of the fluid. With time and distance measured, the terminal velocity may be calculated and viscosity determined from Equation (7.14).

The third general type of viscometer consists of two concentric cylinders with the space between filled by the test liquid. When the inner cylinder is rotated at a fixed angular velocity, shear stresses are set up (see worked example, Chapter 2). These depend on the speed of rotation and on the viscosity of the fluid. Measurements of the applied torque and the speed of rotation can thus be correlated with the viscosity according to:

$$T = K \mu \omega \qquad (7.19)$$

where T = torque, ω = angular velocity and K = coefficient of proportionality.

Again the coefficient of proportionality must be obtained from calibration tests on fluids of known viscosity. For very thick viscous fluids, it is possible to obtain approximate values without using the outer cylinder. In this method the inner cylinder is simply dipped into the fluid and rotated at a fixed angular speed.

WORKED EXAMPLES

Example 7.1: POIV: Poiseuille's apparatus

Flow of water in a Poiseuille's apparatus (see Figure 7.3) gives results

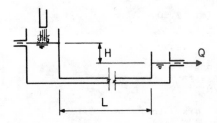

Figure 7.3 Poiseuille's apparatus

shown below. Write a program to calculate mean and maximumm velocities, Reynolds numbers and dynamic viscosities.

Pipe length: 6.18 m, Pipe diameter: 44 mm

Head (mm)	165	205	619	782
Volume (ml)	500	465	500	500
Collection Time (s)	259	207	90	69

```
10 REM "POIV"
20 CLS :KEY OFF
30 PRINT"    This program will calculate viscosity "
40 PRINT"from data obtained using Poiseuille's"
50 PRINT"apparatus":PRINT
60 INPUT"Enter the diameter of the pipe (m)  ";D
70 INPUT"Enter the length of the pipe (m)  "; L
80 INPUT"Please now enter the head (m)   "; H
90 INPUT "Enter the volume collected (ml) "; VOL
100 INPUT "and the collection time (sec) ";T
110 Q=VOL/(T*1*1000000!)
120 GAMMA= 9810
130 RHO = 1000
140 MU = (3.14159*(D^4)*GAMMA*H)/(128*L*Q)
150 R = (Q*D*RHO)/((3.1415/4)*(D^2) * MU)
160 VMAX = (GAMMA*H*((D/2)^2))/(L*4*MU)
170 VMEAN = VMAX/2
180 PRINT:PRINT"Dynamic viscosity (Pa sec) = ";MU
190 PRINT"Reynolds number  = ";R
200 PRINT"Maximum velocity (m/s) = ";VMAX
210 PRINT "Mean velocity (m/s) = ";VMEAN
220 INPUT"Do you wish to continue (Y/N) ";A$
230 IF A$="Y"OR A$="y" THEN CLS:GOTO 80
240 PRINT:PRINT"Program Ended . . . "
250 END
```

```
        This program will calculate viscosity
from data obtained using Poiseuille's
apparatus

Enter the diameter of the pipe (m)  ? 0.0044
Enter the length of the pipe (m)  ? 6.185
Please now enter the head (m)   ? 0.165
```

```
Enter the volume collected (ml) ? 500
and the collection time (sec) ? 259

Dynamic viscosity (Pa sec) =  1.247077E-03
Reynolds number  =  447.9682
Maximum velocity (m/s) =  .2539249
Mean velocity  (m/s) =  .1269625
Do you wish to continue (Y/N) ?  Y

     (The screen will clear at this point)

Please now enter the head (m)   ? .782
Enter the volume collected (ml) ? 500
and the collection time (sec) ? 69

Dynamic viscosity (Pa sec) =  1.574583E-03
Reynolds number  =  1331.759
Maximum velocity (m/s) =  .9531384
Mean velocity  (m/s) =  .4765692
Do you wish to continue (Y/N) ? n

Program Ended . . .
Ok
```

Program notes

(1) Lines 10–50 clear screen, keys off, print heading.
(2) Lines 60–100 enter data.
(3) Lines 110–130 calculate discharge (Q), specify specific weight and density.
(4) Line 140 viscosity from Poiseulle's Equation (7.12).
(5) Line 150 Reynolds number calculation.
(6) Line 160–170 max and mean velocities: V_{max} from Equation (7.9) with $dp/dx = H/L$.
(7) Lines 180–210 print data.
(8) Lines 220–250 routine to permit another calculation or stop.

Example 7.2: PLAV: laminar flow between plates

In Figure 7.4 one plate moves relative to the other as shown.

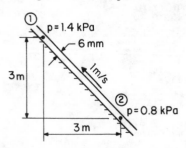

Figure 7.4 Laminar flow between plates

Determine the velocity distribution and the magnitude and location of the maximum velocity. ($\rho = 850 \, \text{kg/m}^3$, $\mu = 0.08 \, \text{Ns/m}^2$)

```
10 REM "PLAV"
20 CLS: KEY OFF
30 PRINT"This program analyses flow between parallel"
35 PRINT"     plates"
40 PRINT:PRINT"Please input upstream pressure     ";
50 INPUT P1
60 PRINT:PRINT"Please input downstream pressure   ";
70 INPUT P2
80 PRINT
90 INPUT"please enter dynamic viscosity (Pa sec) ";MU
100 H1 =850 * 9.810001 * 3: REM 'head one'
110 H2 = 0        :REM 'head two'
120 PTERM1 = P1+H1
130 PTERM2 = P2+H2
140 DELP = PTERM2 - PTERM1
150 DELL = 3 * (2^.5)
160 RATIO = DELP / DELL
170 U = -1
180 AS= .006
190 MU = .08
200 A = (U/AS) -(AS/(2*MU)) * RATIO
210 B=RATIO/(2*MU)
220 PRINT:PRINT"Velocity Distribution is given by:"
230 PRINT"        U = ";A;"Y + ";B;"Y^2"
240 PRINT
250 DIST = -(A/(2*B))
260 PRINT"Maximum velocity occurs at Y = ";DIST
270 UMAX=A*DIST+B*DIST*DIST
280 PRINT:PRINT"Maximum velocity = ";UMAX
290 REM For Q integrate U (from line 210 )
295 REM     with Y from 0 to AS
300 Q=A*AS*AS/2+B*AS*AS*AS/3
305 PRINT
310 PRINT"Flow (per metre of width) = ";Q
320 PRINT"-Negative means up, Positive means down-"
330 REM For du/dy differentiate U (from line 210 )
335 REM     with Y= AS
340 DUDY=A+B*2*AS
350 TAU = MU*DUDY
360 PRINT:PRINT"Shear Stress = ";TAU
370 PRINT:PRINT"Program Ended..."
380 END
```

```
This program analyses flow between parallel
    plates

Please input upstream pressure     ? 1400

Please input downstream pressure   ? 800

please enter dynamic viscosity (Pa sec) ? 0.08

Velocity Distribution is given by:
        U =  59.74456 Y + -37735.2 Y^2
```

```
Maximum velocity occurs at Y =   7.916289E-04

Maximum velocity =   2.364776E-02

Flow (per metre of width) = -1.641533E-03
-Negative means up, Positive means down-

Shear Stress = -31.44623

Program Ended...
Ok
```

Program notes

(1) Lines 10–35 clear screen, keys off, print headings.
(2) Lines 35–90 enter pressures at 1 and 2 (*Pa*) and dynamic viscosity (Ns/m^2).
(3) Lines 100–130 calculate ($\rho gh + P$).
(4) Lines 140–160 calculate dP/dl.
(5) Lines 170–190 specification of velocity, distance between plates and viscosity.
(6) Lines 200–230 velocity from Equation (7.6).
(7) Lines 240–280 location and magnitude of maximum velocity by differentiating velocity equation (line 230) and setting $du/dy = 0$.
(8) Lines 290–320 discharge by integrating velocity equation (line 230).
(9) lines 330–380 shear stress from Equation (2.10). Stop execution.

Example 7.3: SEDV: settling tank

A program to assist in the design of settling tanks would determine (1) the diameter of particles, 100 per cent of which would settle out and (2) the proportion of particles of a given size which would settle, both for a given specific gravity, specified surface area and specified flow. Write such a program and test it with a flow of 20 l/s through a tank with a surface area of 5 m². Investigate also the settling characteristics of particles of diameter 0.04 mm. Take $S = 2.6$ and $v = 10^{-6}$ m²/s.

```
5 REM "SEDV"
10 CLS:KEY OFF
20 S= 2.6 :REM specific gravity
30 PRINT:PRINT"This program will assist in the design"
40 PRINT"of sedimentation tanks by :"
50 PRINT"       1) Determining the particle diameter for"
60 PRINT"          which there would be 100% settlement."
70 PRINT"       2) Determining the percentage settlement"
80 PRINT"          for particles of specified diameter."
85 PRINT:PRINT
```

```
90 PRINT"Specify which option (1 or 2) you require:   ";
100 INPUT A
110 IF A<>1 AND A<>2 THEN GOTO 10
120 CLS:INPUT"Enter the spec grav of the particles";S
130 PRINT:PRINT"Please enter the flow (m3/sec)      ";
140 INPUT Q
150 PRINT:PRINT"Please enter the surface area  (m2)   ";
160 INPUT AS
170 IF A=2 THEN GOTO 240
180 VS=Q/AS
190 D=SQR((VS*18)/(9.810001*(S-1)*1000000!))
200 D=D*1000
210 D=INT(D*1000)/1000
215 PRINT
220 PRINT"Diameter for 100% settlement   =";D;"  (mm)"
225 PRINT
230 GOTO 290
240 PRINT"Please enter the particle diameter  (m)   ";
250 INPUT DP
260 V1=(9.810001*(DP^2)*(S-1))/(18*(1*10^-6))
270 P=(V1/(Q/AS))*100
280 PRINT"Portion of particles which settle  = ";P;"%"
285 PRINT:PRINT
290 PRINT"do you want to calculate another? (y/n)   ";
300 INPUT A$
310 IF A$="y"OR A$="Y" THEN GOTO 10
320 PRINT:PRINT"program ended..."
330 END
```

This program will assist in the design
of sedimentation tanks by :
        1) Determining the particle diameter for
           which there would be 100% settlement.
        2) Determining the percentage settlement
           for particles of specified diameter.

Specify which option (1 or 2) you require: ? 1
Enter the spec grav of the particles? 2.6

Please enter the flow (m3/sec)   ? 0.02

Please enter the surface area (m2)   ? 5

Diameter for 100% settlement   = .067   (mm)

do you want to calculate another? (y/n)   ? y

This program will assist in the design
of sedimentation tanks by :
        1) Determining the particle diameter for
           which there would be 100% settlement.
        2) Determining the percentage settlement
           for particles of specified diameter.

Specify which option (1 or 2) you require: ? 2
Enter the spec grav of the particles? 2.6

```
Please enter the flow (m3/sec)      ? 0.02

Please enter the surface area   (m2)  ? 5
Please enter the particle diameter   (m)   ? 0.00004
Portion of particles which settle  =   34.88 %

do you want to calculate another? (y/n)   ? n

program ended...
Ok
```

Program notes

(1) In a rectangular settling tank (width W, depth H and length L) all particles with a settling velocity equal to or greater than V_s as shown in Figure 7.5 will settle out. From similar triangles:

$$\frac{V}{V_s} = \frac{L}{H}$$ where V = average through flow velocity

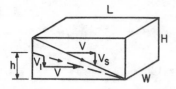

Figure 7.5 Rectangular settling tank

$$\therefore \quad V_s = \frac{VH}{L} = \frac{QH}{WHL} = \frac{Q}{A_s}$$ where A_s = surface area

For particles with settling velocity $V_1 > Vs$

$$P\% = \frac{h}{H} \times 100 = 100 \, V_1 A_s / Q$$

The settling velocities, V_s and V_1 may be obtained from Stoke's Law.
(2) Lines 5–40 clear screen, keys off, print heading.
(3) Lines 50–110 option menu.
(4) Lines 120–160 enter flow and surface area.
(5) Line 170 go to second choice routine.
(6) Lines 180–220 calculate V_s as above and diameter from Stoke's Law.
(7) Line 230 jump to another calculation choice.
(8) Lines 240–280 calculation of P as above, V_1 being given by Stoke's Law.
(9) Lines 285–310 routine for another calculation.

PROBLEMS

(7.1) Rewrite worked Example 7.1, POIV, to record the results of up to ten laboratory experiments, to print details of flow, Reynolds numbers and viscosities and to calculate the average value of viscosity based on the full set of experiments.

(7.2) Fluid of density $1200 \, kg/m^3$ and viscosity $0.95 \, Ns/m^2$ runs between two stationary, infinite, parallel plates $1.5 \, cm$ apart. Write a program which would give results assisting the generation of a graph showing pressure drop per unit length against discharge per unit width.

(7.3) Write a program to determine the terminal velocity of steel spheres ($S = 7.8$; $0.01 \, cm <$ diameter $< 2.0 \, cm$) falling through fluids of specific gravity 1.2 and viscosities ranging from $.007 \, Ns/m^2$ to $50 \, Ns/m^2$. Print results in tabular form.

(7.4) Glycerine ($\mu = 1.5 \, Ns/m^2$, $S = 1.26$) flows in a $20 \, cm$ diameter pipe with a velocity of $30 \, cm/s$. Develop a program to assist with plotting the velocity distribution across the cross section. Write the program to cater for other sizes of pipe and other fluids with different velocities.

(7.5) The program SEDV contains a number of GOTO statements. Rewrite the program replacing all GOTOs by WHILE-WEND loops.

Chapter 8

Flow measurement

8.1 Velocity measurement

Current meters (for water) or anemometers (for air) consist basically of propellers or cups which rotate about an axis when immersed in the flow. The propeller, of course, rotates about an axis parallel to the flow direction while the cups rotate about an axis perpendicular to the flow. In both cases the velocity of the flow is correlated with the speed of rotation. Current speeds are read from a gauge or from an audible count provided through earphones. Both systems may be calibrated by towing the meters through liquid or air, as appropriate, at known speeds. Current meters are typically used for the measurement of velocity in open channels and vary significantly in size depending on the particular application. For laboratory use, meters are typically small with propellers around 10 to 20 mm in diameter. Much larger propellers and cup type devices would be used in natural channels where they might be held in position by a fixed rod used while wading. In larger rivers, where wading is impossible, the meter would be suspended from a boat or a bridge or from a cable spanning the river.

In both cases it is extremely important to have the axis of rotation correctly aligned with the flow. Significant errors may result from small alignment errors.

The pitot tube works on the principal shown in Figure 8.1. With the tube bent through 90° and aligned with the flow direction, fluid comes to rest immediately in front of the tube at a point called the stagnation point. The pressure here, termed the stagnation pressure, may be evaluated using Bernoulli's equation to give:

$$\frac{p}{\gamma} + \frac{V^2}{2g} = \frac{p_s}{\gamma} \tag{8.1}$$

where p_s = stagnation pressure.

The increase in pressure above the static pressure in the pipe causes water in the pitot tube to rise to a height h above the hydraulic grade line and measurement of h then provides a direct measurement of velocity. This may be seen by rearranging Equation (8.1) to give:

$$h = \frac{p_s - p}{\gamma} = \frac{V^2}{2g} \tag{8.2}$$

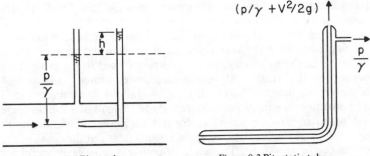

Figure 8.1 Pitot tube   Figure 8.2 Pitostatic tube

Determination of velocity in this manner would require both a pitot tube and a piezometric tube but these may be combined in one instrument called a pitotstatic tube. This is shown in Figure 8.2. Here the central tube measures the total head while the static head is picked up from a series of small holes set around the outside of the outer tube. The two tubes are connected to a differential pressure manometer which gives a direct reading of the head difference. Because of turbulence in the flow approaching the pitotmeter, the fluid particles do not travel in straight parallel paths and a coefficient is necessary to account for the slight errors that this causes. This may be incorporated into Equation (8.2) to give:

$$V = C\left[2g\left(\frac{p_s}{\gamma} - \frac{p}{\gamma}\right)\right]^{1/2} \tag{8.3}$$

Typically the constant C would be determined by experiment.

Many other methods of velocity measurement have been developed. Hot wire anemometers operate essentially by measuring heat transfer to the fluid. The rate at which heat is dissipated from the wire depends on the velocity of the surrounding fluid while the electrical resistance of the wire varies with temperature. Measurement of current through the wire at constant voltage (or voltage at constant current) thus provides a measurement of the velocity of the fluid in which the wire is immersed.

Surface velocities may be estimated from the motion of floats. These can provide crude estimates such as would be obtained by throwing pieces of wood into a river and measuring the distance travelled in given times. Alternatively, quite sophisticated measurements may be made using photographic techniques with flashing lights and long time exposures. Such techniques are often used in hydraulic models.

Simple experimental methods of measuring velocity may be derived using the drag on an object immersed in the flow. Small spheres suspended on thin wires or thread will be deflected by the flow depending on the velocity at the sphere and this can be correlated with the angle at which the wire becomes inclined to the vertical. These might be useful at small scale but are rarely used in the laboratory because of the availability of more accurate methods. In the field however, the drag method is used extensively to determine velocity and direction of subsurface ocean currents. Instead of a small sphere a drag device consisting of a cruciform shape made out of aluminium or wood would be constructed and set, using floats, at the depth of the current to be measured. These are, of course, particularly useful for currents close to the surface and the drogues might, for example, be set about two metres below the surface suspended by a wire cable attached to a small surface float. The floats would then be tracked by radar or theodolite to obtain direction and time of motion.

8.2 Discharge measurement

Nozzles and orifices have been dealt with earlier in Chapter 4 where it was shown that the flow from either could be given as:

$$Q = Cd \, A \, (2gh)^{1/2} \tag{8.4}$$

in which case the flow may be determined by measuring the head h. In the case of nozzles, this head would normally include pressure and velocity head as shown in Figure 5.7.

An orifice may also be used as a means of discharge measurement when it is located in a pipe as shown in Figure 8.3. The application of

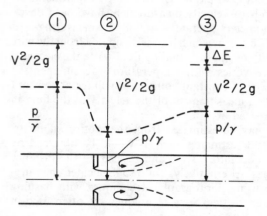

Figure 8.3 Orifice meter

Bernoulli's equation from the undisturbed flow upstream of the orifice to a point in the pipe just downstream leads, in the assumption of no losses, to:

$$\frac{p_1}{\gamma} + \frac{V_1^2}{2g} + Z_1 = \frac{p_2}{\gamma} + \frac{V_2^2}{2g} + Z_2 \qquad (8.5)$$

But $V_2 = V_1 (D_1/D_2)^2$

and $Q = V_1 \frac{\pi}{4} D_1^2$.

Therefore by substitution:

$$Q = C \frac{\pi}{4} D_1^2 \left\{ \frac{2g \left(\frac{p_1 - p_2}{\gamma} \right) + Z_1 - Z_2}{(D_1/D_2)^4 - 1} \right\}^{1/2} \qquad (8.6)$$

where C is a coefficient necessary to account for the contraction of the jet and the losses which occur downstream of the orifice.

The orifice meter has the considerable advantage that it is relatively cheap and easy to install. However, significant energy losses are associated with orifice plates and in northern climates there is an added disadvantage of ice accumulation. Typically, coefficients vary from about 0.6 to about 0.8.

The venturi meter shown in Figure 8.4 overcomes many of the disadvantages associated with orifice meters. The careful tapering of the entrance to the throat combined with a slow flare downstream reduces head losses thus giving meter coefficients much closer to unity. Because of the similarity of the flow through an orifice meter and a venturi meter, Equation (8.6) applies to both but the careful tapering upstream of the throat and the slow flare in the pipe down-

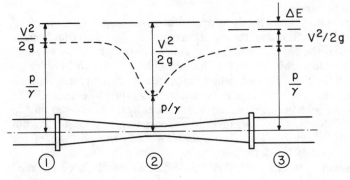

Figure 8.4 Venturi meter

stream lead to much better values of the meter coefficient. Typically for an orifice meter, the value of the coefficient might range from about .95 to about .99 depending on the diameter ratios.

In open channels, the discharge can be measured using a series of velocity traverses in which the cross section of the river or channel is divided into a number of sections and the velocity is measured at a series of points along a vertical line in the centre of each section. The average velocity may then be evaluated for each section and the discharge calculated by a summation, i.e.

$$Q = V_1 A_1 + V_2 A_2 + V_3 A_3 + \ldots \tag{8.7}$$

This method was described in detail in *BASIC Hydrology*.

Discharge measurements by the injection of some tracer which will mix with the flow are recommended only in cases where other methods are difficult or unsatisfactory, for example, in rapidly flowing highly turbulent mountain streams. The salt injection method, as it is commonly called, consists of injecting a tracer at some known concentration into the stream and measuring the resulting concentration in the channel some distance downstream where the tracer has thoroughly mixed with the channel water. Based on the principle of continuity of the tracer, the flow in the river is given by:

$$Q = \frac{Q_{tr} C_i}{C_s} \tag{8.8}$$

where Q = flow in channel, Q_{tr} = flow of tracer at injection point, C_i = tracer concentration at injection point and C_s = tracer concentration at sampling point.

This equation assumes a continuous injection of tracer and also assumes that the tracer is thoroughly and completely mixed with the channel water at the sampling point. It is important therefore to locate the sample point sufficiently far downstream to ensure thorough mixing.

Other injection methods have also been suggested. One fairly sophisticated method uses the injection of air through a pipe laid across the bed of the channel and discharging air through a series of small holes spread evenly across the channel width. As the air bubbles lift through the fluid, they essentially integrate the velocity during their vertical rise. The discharge in the channel can then (theoretically) be correlated with the shape of the profile made by the bubbles at the top of the water column. This method appears to give good results but depends, of course, on being able to identify the position of the bubbles as they break the surface.

In small open channels, flow measurement may be accomplished by the use of sharp edged weirs and notches. These consist of a plate inserted in the channel so that all of the flow passes over the plate.

Typical shapes are shown in Figure 8.5. The flow over the weir or notch may be analysed using a combination of energy and continuity considerations or perhaps more conveniently by the methods discussed in Chapter 9. If the velocity of approach is ignored, these analyses result for a Vee notch weir, in:

$$Q = K \, H^{5/2} \tag{8.9}$$

and for a rectangular weir, in:

$$Q = K \, L \, H^{3/2} \tag{8.10}$$

where H = head over weir and K = a coefficient depending on weir shape.

In a Vee notch weir, for example:

$$K = C_d \frac{8}{15} (2g)^{1/2} \tan \phi$$

where ϕ = half angle of notch.

Many other types of flow measurement device also exist. Rotameters rely on the drag of a heavier than water 'float' located inside a vertical tapered tube for their operation. As the flow increases, the drag on the 'float' becomes greater and it rises in the tapered tube. The magnitude of the flow can be correlated with the position of the 'float'. Positive displacement meters, often used for water supply, operate with vanes which are spring loaded to maintain continuous contact with the meter casing. The rotation of the vanes then registers the total flow through the meter with an accuracy of around 0.5%. Magnetic flow meters set up a magnetic field across the flow path and, since the fluid is a conductor moving in the field, it will set up an induced voltage which may be measured and correlated with the discharge. Laser-anemometers focus a laser beam on a small point in the fluid. The particles contained by the fluid cause the light to be scattered with a frequency shift which is directly proportional to the flow velocity. Rather sophisticated electronic measuring devices however are required and this method tends to be expensive.

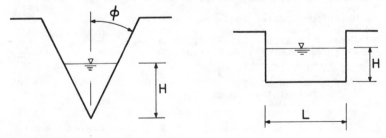

Figure 8.5 Sharp crested weirs

WORKED EXAMPLES

Example 8.1: RECE: rectangular weir errors

The equation for flow per unit width over a rectangular weir (Figure 8.6) is:

$$q = Q/L = \frac{2}{3}(2g)^{1/2}\left[\left(H + \frac{V^2}{2g}\right)^{3/2} - \left(\frac{V^2}{2g}\right)^{3/2}\right]$$ (8.11)

which is often simplified to:

$$q = \frac{2}{3}(2g)^{1/2}\, H^{3/2}$$ (8.12)

Write a program to obtain a solution for Equation (8.11) by trial and error using Equation (8.12) as a first estimate. Then calculate the error caused by using equation (8.12) instead of Equation (8.11).

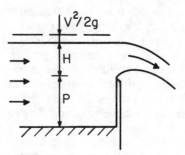

Figure 8.6 Rectangular weir

```
10 REM "RECE"
20 CLS : KEY OFF
30 PRINT"    This program will make successive "
35 PRINT"approximations of the flow over a"
40 PRINT"rectangular weir. The user must provide"
45 PRINT"the acceptable level of error in decimal form"
50 PRINT
60 PRINT:INPUT"Enter the acceptable error value   "; E
70 INPUT"Please enter the head over weir    "; H
80 INPUT "Please enter the height of weir    ";P
90 Q1= (2/3)*SQR(2*9.810001)*(H^1.5)
100 QA=Q1
110 I=1
120 WHILE ABS(QA-QC)>E
130     IF I<>1 THEN QA=QC
140     V=QA/(H+P)
```

```
150      QC=(2/3)*SQR(2*9.810001)
155      T1=(V*V/(2*9.810001))^1.5
160      QC=QC*((H+V*V/(2*9.810001))^1.5 -T1^1.5)
170      I=I+1
180 WEND
190 PRINT:PRINT"Discharge  = ";QC
200 PRINT"Approx Discharge  = ";Q1
210 PRINT "Height of weir  = ";P
220 PRINT "Error  = ";(QC-QA)/QC;" %"
230 PRINT:PRINT"Program Ended..."
240 END



    This program will make successive
approximations of the flow over a
rectangular weir. The user must provide
the acceptable level of error in decimal form



Enter the acceptable error value  ? .0001
Please enter the head over weir   ? .4
Please enter the height of weir   ? .6


Discharge  =  .8510306
Approx Discharge  =  .7470476
Height of weir  =  .6
Error  =  7.44506E-05  %


Program Ended...
Ok
```

Program notes

(1) Lines 10–50 clear screen, keys off, print headings.
(2) Lines 60–80 input data.
(3) Lines 90–110 calculate approximate discharge, set counter.
(4) Lines 120–180 loop to calculate exact discharge with error test.
(5) Lines 190–240 print results and stop execution.

Example 8.2: VEEW: vee notch weir calibration

A laboratory system for calibrating a vee notch weir measures the flow using a weigh scale and time clock. Results of ten experiments are used to determine the value of the discharge coefficient from Equation (8.9) and (8.10). Write a program to record experimental results and to calculate the average value of the discharge coefficient. Test the program with the following results.

Experiment	T_1 (s)	T_2	W_1 (N)	W_2	H (m)
1	10	40	0	132.6	0.05
2	40	80	0	566.2	0.08
3	80	120	0	1017.1	0.10
4	120	172	0	2037	0.12
5	172	200	0	1634	0.14
6	200	225	0	1731	0.15
7	225	246	0	3026	0.20
8	246	266	0	3645	0.22
9	266	294	0	6907	0.25
10	294	312	0	6980	0.30

```
10 REM "VEEW"
20 CLS : KEY OFF
30 PRINT"   This program calculates a discharge"
35 PRINT"coefficient based on the results of ten"
40 PRINT"experiments with a Vee notch weir. The"
45 PRINT"user must supply the notch angle, and ten"
50 PRINT"values of time, weight and head for each of"
55 PRINT"the ten experiments. Initial and final "
60 PRINT"values are requested to accomodate "
65 PRINT"cumulative results from consecutive users"
75 PRINT
80 PRINT"Please input the Vee notch half angle (deg)   ";
90 INPUT THETA
100 THETA =THETA*3.14159/180
105 PRINT
110 PRINT"Press return to begin entering results....";
120 INPUT DUMMY
125 REM DUMMYis meaningless, just waiting for [return]
130 FOR I= 1 TO 10
140    CLS:PRINT:PRINT "Experiment Number:    ";I
150    PRINT:INPUT"Please input initial time   ";T1(I)
160    PRINT:INPUT"Please input final time    ";T2(I)
170    PRINT:INPUT"Please input initial weight   ";W1(I)
180    PRINT:INPUT"Please input final weight    ";W2(I)
190    PRINT:INPUT "Enter head during test    ";H(I)
200    PRINT:INPUT "Enter 1 if correct, 0 to change   ";Z
210    IF Z=0 THEN GOTO 140
220    IF Z=1 THEN GOTO 240
230    GOTO 200
240    Q(I)=(W2(I)-W1(I))/(9810*(T2(I)-T1(I)))
245    TERM=((2*9.810001)^.5)*(H(I)^2.5)
250    C(I)=Q(I)/((8/15)*TAN(THETA)*TERM)
260    Q(I)=INT(Q(I)*10000)/10000
270    C(I)=INT(C(I)*1000)/1000
280 NEXT I
290 TOT=0
300 FOR I=1 TO 10
310 TOT= TOT+C(I)
320 NEXT I
330 CA=TOT/10
340 CLS
350 PRINT:PRINT
```

```
360 PRINT:PRINT"Average Discharge Coefficient  = ";CA
370 PRINT:PRINT
380 PRINT" EXP   T1   T2   W1   W2   H    Q    C "
390 PRINT"_____"
400 FOR I=1 TO 10
410    PRINT I" ";T1(I)" ";T2(I)" ";W1(I)" ";
415    PRINT W2(I)" ";H(I)" ";Q(I)" ";C(I)
420 NEXT I
430 END
```

This program calculates a discharge
coefficient based on the results of ten
experiments with a Vee notch weir. The
user must supply the notch angle, and ten
values of time, weight and head for each of
the ten experiments. Initial and final
values are requested to accomodate
cumulative results from consecutive users

Please input the Vee notch half angle (deg) ? 30

Press return to begin entering results....?

Experiment Number: 1

Please input initial time   ? 10

Please input final time   ? 40

Please input initial weight   ? 0

Please input final weight   ? 132.6

Enter head during test   ? 0.05

Enter 1 if correct, 0 to change ? 1

Continue to enter results until all 10 expts are filed
  Pressing return will then give

Average Discharge Coefficient = .5925

EXP	T1	T2	W1	W2	H	Q	C
1	10	40	0	132.6	.05	.0004	.59
2	40	80	0	566.2	.08	.0014	.584
3	80	120	0	1017.1	.1	.0025	.6
4	120	172	0	2037	.12	.0039	.586
5	172	200	0	1634	.14	.0059	.594
6	200	225	0	1731	.15	.007	.593
7	225	246	0	3026	.2	.0146	.602
8	246	266	0	3645	.22	.0185	.6
9	266	294	0	6907	.25	.0251	.589
10	294	312	0	6980	.3	.0395	.587

Ok

Program notes

(1) Lines 10–75 clear screen, keys off and print heading.
(2) Lines 80–230 input data and check each entry.
(3) Lines 240–280 calculate discharge from weight and coefficient from Equation (8.10).
(4) Lines 290–360 calculate and print average coefficient.
(5) Lines 370–430 print experimental data and stop.

Example 8.3: TUBV: velocity in a tube

Two methods exist for calculating the flow in a cylindrical pipe based on velocity measurements. In one the flow is given by the product of the cross sectional area and the average velocity. In the other the square of the velocity is plotted against radius and the flow is given by the product of π and the area under the $V–R^2$ curve. Write a program to compare the two methods and check it with the following data.

Radius (m)	0.0	0.1	0.3	0.45	0.50
Velocity (m/s)	0.51	0.48	0.31	0.09	0.0

```
10 REM "TUBV"
15 DIM  R(5),V(5)
20 CLS: KEY OFF
30 PRINT"This program will calculate the flow"
35 PRINT"in a cylindrical pipe, given five sets"
40 PRINT"of data using both the 'area' method"
50 PRINT"and the average velocity method"
60 PRINT:PRINT
70 FOR I =1 TO 5
80     PRINT"Enter radius no ";I;" (M) ";
90     INPUT R(I)
100    PRINT"Enter velocity no ";I;"  (m/s) ";
110    INPUT V(I)
130 NEXT I
140 CLS
150 PRINT
160 PRINT"Radius (m)  Velocity (m/s)"
170 PRINT"--------------------------"
180 FOR I = 1 TO 5
190 PRINT R(I), V(I)
200 NEXT I
210 PRINT:PRINT"Is this correct? (Y/N)  ";
220 INPUT A$
230 IF A$="Y" OR A$="y" THEN GOTO 260
240 IF A$="N" OR A$="n" THEN GOTO  60
250 GOTO 210
260 REM  Area calculation
270 FOR I = 1 TO 4
275    TERM2=(((R(I+1)^2)-(R(I)^2)))
```

```
280     AREA = AREA+((V(I)+V(I+1))/2)*TERM2
285 NEXT I
290 QA=3.14159*AREA
300 REM Average Velocity Calculation
305 TERM1=(3.14159/4)*((2*R(5))^2)
310 QV = ((2*V(2)+2*V(3)+2*V(4))/6)*TERM1
320 PRINT"Discharge by area   (m3/s) =  ";QA
330 PRINT"Discharge by av vel (m3/s) =  ";QV
340 PRINT:PRINT "Another Calculation? (Y/N) ";
350 INPUT B$
360 IF B$="y" OR B$="Y" THEN GOTO 10
370 IF B$="n" OR B$="N" THEN GOTO 400
380 GOTO 340
400 PRINT:PRINT "Program Ended..."
410 END
```

```
This program will calculate the flow
in a cylindrical pipe, given five sets
of data using both the 'area' method
and the average velocity method


Enter radius no  1  (M) ? 0
Enter velocity no. 1    (m/s) ? 0.51
Enter radius no  2  (M) ? 0.1
Enter velocity no  2    (m/s) ? 0.48
Enter radius no  3  (M) ? 0.3
Enter velocity no  3    (m/s) ? 0.31
Enter radius no  4  (M) ? 0.45
Enter velocity no  4    (m/s) ? 0.09
Enter radius no  5  (M) ? 0.50
Enter velocity no  5    (m/s) ? 0

Radius (m)   Velocity (m/s)
--------------------------
0                 .51
.1                .48
.3                .31
.45               9.000001E-02
.5                0

Is this correct? (Y/N)  ? y
Discharge by area   (m3/s) =   .192226
Discharge by av vel (m3/s)  =   .2303833

Another Calculation? (Y/N) ? N

Program Ended...
Ok
```

Program notes

(1) Lines 10–60 clear screen, keys off, print headings.
(2) Lines 70–250 enter data with check for correct entry.
(3) Lines 260–285 area of V–R^2 curve by Simpson's rule.
(4) Line 290 discharge calculation.

(5) Lines 300–310 discharge by average velocity.
(6) Lines 320–400 print data and stop unless more calculations.

Example 8.4: EFAC: expansion factor

Compressible gas flow calculations involving high velocities at a pipe constriction are often performed using continuity and energy principles for incompressible flow together with an expansion factor to take account of the gas compression. This factor, Y, is given by:

$$Y = \sqrt{\frac{1 - \left(\frac{A_2}{A_1}\right)^2 \frac{k}{k-1} \left(\frac{p_2}{p_1}\right)^{\frac{2}{k}} \left[1 - \left(\frac{p_2}{p_1}\right)^{\frac{k-1}{k}}\right]}{1 - \left(\frac{A_2}{A_1}\right)^2 \left(\frac{p_2}{p_1}\right)^{\frac{2}{k}} 1 - \frac{p_2}{p_1}}}$$

where A = area, p = pressure and k = ratio of specific heats.

Write a program using nested loops to print a table of values of Y, for $A_2/A_1 = 0.2$, $1.4 < k < 1.3$ and $0.95 < P_2/P_1 < 0.75$.

```
10 REM "EFAC"
20 CLS : KEY OFF
30 PRINT"  This program will develop a table "
35 PRINT"of expansion factors for varying values"
40 PRINT"    of (A2/A1), k2, and (p2/p1). "
50 PRINT:PRINT"A2/A1","k","p2/p1","Exp. Factor"
60 PRINT:PRINT
70    FOR A =.15  TO  .6 STEP .15
80       FOR K = 1.4 TO 1.2 STEP -.05
90          FOR P = .95 TO .74 STEP -.05
100            E=(1-A*A)/(1-A*A*P^(2/K))
105            TERM=(1-P^((K-1)/K))/(1-P)
110            D=K/(K-1)*P^(2/K)*TERM
120            F=SQR(E*D)
130            F=INT(F*1000)/1000
140            PRINT A, K, P, F
150          NEXT P
160       NEXT K
170    NEXT A
180 PRINT:PRINT"Program Ended..."
190 END
```

      This program will develop a table
of expansion factors for varying values
      of (A2/A1), k2, and (p2/p1).

A2/A1	k	p2/p1	Exp. Factor
.15	1.4	.95	.972
.15	1.4	.9	.943
.15	1.4	.8499999	.913
.15	1.4	.8	.883
.15	1.4	.75	.852
.15	1.35	.95	.971
.15	1.35	.9	.941

```
*** This output was stopped by CTRL-BREAK. Output
    will normally continue until***
```

.6	1.25	.75	.764
.6	1.2	.95	.946
.6	1.2	.9	.896
.6	1.2	.8499999	.848
.6	1.2	.8	.801
.6	1.2	.75	.756

```
Program Ended...
Ok
```

Program notes

(3) Lines 10–60 clear screen, keys off, print headings.
(2) Lines 70–170 calculate and print factor (lines 100–140) in nested loop set up to vary area ratio, k, and pressure ratio.
(3) Lines 180–190 stop execution.

PROBLEMS

(8.1) Write a program which will permit calculation of the flow over a vee notch weir or rectangular sharp crested weir, the choice of weir to be determined by the user. Allow for different notch angles and different crest lengths.

(8.2) Write a program to be used for developing a calibration chart for a venturi meter. The program should determine the coefficient of discharge based on a number of experiments and should then use that coefficient to obtain a numerical relationship between discharge and head.

(8.3) The channel cross section shown in Figure 4.4 may be divided into three sections by vertical lines drawn through the point (− 8.4) and through the origin. Assuming that the average velocities in the three sections are 0.5 m/s, 0.6 m/s and 0.55 m/s respectively from left to right, calculate the total discharge. Use the program XSECT as a basis for calculation and modify or extend it as necessary.

(8.4) Develop a general program along the lines of problem (8.3) to handle a channel which can be divided into ten sections.

(8.5) Flow measurements in a duct 500 mm square are made by traversing a pitot tube across the duct. Velocity readings are obtained as given below. Assuming the flow is symmetrical about the centre line of the duct, determine the discharge.

Distance from centre line (mm)	–	50	100	150	200	220	250
Velocity (m/s)	0.5	0.49	0.47	0.42	0.33	0.19	0

(8.6) Modify the program VEEW to use 'Print using' statements in order to tidy up the tabular output.

Chapter 9
Dimensional analysis and similarity

9.1 General

Dimensional analysis and similarity methods are methods of partial analysis based essentially on the principle of dimensional homogeneity. Both are partial in the sense that the analysis is incomplete. For example, in very simple terms, a partial analysis would lead to the idea that the hypotenuse of a right angle triangle could be related to other variables of the triangle by the equation:

$$D/L = \phi \, (B/L) \tag{9.1}$$

where D = hypotenuse, L = length, B = breadth and ϕ = function of.

This equation says that the ratio D/L is a function of, or is dependent on, the ratio B/L and the analysis is incomplete because no information is provided regarding the nature of the function. By partial analysis, it is impossible to show that the hypotenuse is equal to the square root of the sum of the squares. Analyses of this type are important and useful in fluid situations where the flow, or the boundaries of the flow, are so complex as to prevent a complete mathematical analysis. The advantage of the partial analysis is threefold. It simplifies the problem, and in Equation (9.1) it can be seen that a problem involving three variables D, L and B has been simplified to one involving only two parameters, namely the two ratios. Partial analysis also provides a guide to experimentation. Again Equation (9.1) can be used as an example. Here, it is clearly necessary to vary the ratio B/L in order to conduct experiments on triangles but it would be unnecessary to vary B and L individually although this should be done for practical reasons. Thirdly, the analysis provides a guide to the design and operation of models. Any model of a triangle can be constructed by maintaining the same value of B/L in the model and in the full scale triangle, called the prototype. In that way, knowledge about the hypotenuse of a full scale triangle could be obtained by constructing a model, measuring the hypotenuse in the model and scaling up to the prototype size knowing that the ratio D/L must be the same in both model and prototype, i.e.

$$D_p = D_m \, (L_p/L_m) \tag{9.2}$$

119

where suffices m and p refer to model and prototype respectively.

The example given using a triangle is obviously trivial but the comments made are all applicable to much more complex situations. In particular, the simplification achieved is much more significant for complex problems involving a large number of variables.

In this chapter three methods of dimensional analysis will be considered and these will be followed by a discussion of similarity theory and the application to models.

9.2 Raleigh's indicial method

The method developed by Lord Raleigh follows directly from the principle of dimensional homogeneity. In any correct equation, each term of the equation must have the same dimensions. Any variable or combination of variables has dimensions which can be expressed in terms of mass M (or force F), length L and time T. Table 9.1 provides a list of common fluid variables and their associated dimensions.

All equations are of the form:

$$A = B + C + D + \ldots \tag{9.3}$$

where A, B, C etc. represent terms in the equation.

Dimensional homogeneity means simply that the dimensions of A must be the same as the dimensions of B and of the dimensions of C etc. throughout the equation. Thus, while it is possible to say that 2 + 2 = 4, such an equation is only valid if the dimensions of each part of the equation are the same. Clearly it is not correct to say that 2 oranges plus 2 apples = 4 pears. In terms of the 'fundamental' dimensions mass, length and time, this means that the exponent of mass, the exponent of time and the exponent of length must be the same in each term in the equation. This feature of dimensional homogeneity is used directly in the indicial method which may be illustrated by considering the flow over a veenotch weir. On the assumption that the discharge Q depends only on gravitational acceleration g, head over the weir H, the weir angle θ and on the viscosity μ and density ρ, a basic functional equation can be written among these variables. Thus:

$$\phi\,(Q, g, H, \theta, \mu, \rho) = 0 \tag{9.4}$$

From this it follows that:

$$K\,[Q]^{a}\,[g]^{b}\,[H]^{c}\,[\theta]^{d}\,[\mu]^{e}\,[\rho]^{f} = M^{\circ}L^{\circ}T^{\circ} \tag{9.5}$$

where [ ] refers to the dimensions involved K is a dimensionless constant and a, b, c etc. are unknown exponents.

If the equation is now written in terms of the dimensions of the variables, then:

$$\left(\frac{L^3}{T}\right)^a \left(\frac{L}{T^2}\right)^b (L)^c (1)^d \left(\frac{M}{LT}\right)^e \left(\frac{M}{L^3}\right)^f = M^\circ L^\circ T^\circ \tag{9.6}$$

Because the exponents of mass, length and time must be the same on both sides, Equation (9.6) can now be used to generate three different equations as follows:

$$L: \quad 3a + b + c - e - 3f = 0 \tag{9.7}$$

$$M: \quad e + f = 0 \tag{9.8}$$

$$T: \quad -a - 2b - e = 0 \tag{9.9}$$

With three equations it is now possible to solve for three of the unknown exponents in terms of the other. Thus

$$b = \frac{-e - a}{2} \tag{9.10}$$

$$f = -e \tag{9.11}$$

$$c = -\frac{3}{2}e - \frac{5}{2}a \tag{9.12}$$

Substituting back into Equation (9.5) then gives

$$K \left(\frac{Q}{g^{1/2} H^{5/2}}\right)^a \times \left(\frac{\rho g^{1/2} H^{3/2}}{\mu}\right)^{-e} \times \theta^d = 0 \tag{9.13}$$

Because k, a, e and d are indeterminate, numerical values cannot be assigned. Instead Equation (9.13) may be written in functional form as:

$$\frac{Q}{g^{1/2} H^{5/2}} = \phi\left[\left(\frac{\rho g^{1/2} H^{3/2}}{\mu}\right), \theta\right] \tag{9.14}$$

If the flow over the notch is fully turbulent, then the viscosity of the fluid will be unimportant. Viscosity may be then dropped from the analysis to give:

$$\frac{Q}{g^{1/2} H^{5/2}} = \phi(\theta) \tag{9.15}$$

or, for a particular notch angle:

$$Q = Cg^{1/2} H^{5/2} \tag{9,16}$$

which is the same as Equation (8.9).

Finally, it may be noted that the problem which was initially specified in terms of six variables (Equation (9.4)) has been simplified through partial analysis to one involving three parameters (Equation (9.14)).

9.3 Buckingham π theorem

Buckingham's theorem specifies the number of parameters needed to define a particular phenomenon based on the number of variables present and the number of dimensions associated with the problem. In general, if there are m variables and n dimensions, then $m-n$ parameters are required. In the case of the vee notch discussed earlier, Equation (9.4) shows that six variables are present. There are three pertinent dimensions, mass, length and time, so Buckingham's theorem leads to the idea that there must be three parameters (as shown in Equation (9.14)). However, it is important to recognize that there may be less than three fundamental dimensions. The problem discussed at the beginning of the chapter deals with the geometry of triangles in which the only relative dimension is length. Thus, with three variables and one dimension there must be two parameters (see Equation 9.1)).

Buckingham also suggested a method of approach in which each dimensionless parameter (a π term) is derived independently from the other. With m dimensions it is necessary to select m variables and combine these with each of the others one at a time in order to form a π term. It is important however, to ensure that each of the dimensions pertinent to the problem are contained in the m variables chosen.

In the case of the vee notch discussed earlier, three variables containing mass, length and time must be chosen and these are then used as repeating variables; i.e. these variables appear in each of the π terms. Choosing, for example, ρ, g and H and combining these one at a time with each of the other variables, i.e. μ, Q and θ would lead to:

$$\pi_1 = \rho^a\, g^b\, H^c\, \mu^d \tag{9.17}$$

$$\pi_2 = \rho^a\, g^b\, H^c\, Q^d \tag{9.18}$$

$$\pi_3 = \rho^a\, g^b\, H^c\, \theta^d \tag{9.19}$$

The exponents in each of these equations may be calculated using the method outlined in the indicial equation. Thus for π_3:

$$(MLT)^o = \left(\frac{M}{L^3}\right)^a \left(\frac{L}{T^2}\right)^b (L)^c\, (1)^d \tag{9.20}$$

Equating exponents on the left and right hand side of Equation (9.20) quickly shows that a, b and c are all 0 while d is indeterminate.

Thus, the third π term must be θ. Similarly, π_1 can be shown to be the first term inside the brackets of Equation (9.14) while π_2 is given by:

$$\pi_2 = \frac{gH^5}{Q^2} \qquad (9.21)$$

It will be noted that the format of π_2 is different in Equation (9.21) from the format of the first term in Equation (9.14). This occurs because a partial analysis does not define the function and there are an infinite variety of correct solutions. For example, if A depends on B then equally A depends on any constant multiplied by B raised to any power. This technique, known as compounding, may be used to combine parameters in order to obtain the most convenient solution.

9.4 Matrix methods

As an alternative to Equation (9.5), the relationship among the various indices may be determined using matrix algebra. The dimensions of each of the pertinent variables may be displayed in matrix form as shown in Equation (9.22) and the matrix may then be solved to give the three Equations (9.7), (9.8) and (9.9). The original matrix is shown below:

	(a) Q	(b) g	(c) H	(d) θ	(e) μ	(f) ρ		
L	3	1	1	0	-1	-3	0	
M	0	0	0	0	1	1	$= 0$	(9.22)
T	-1	-2	0	0	-1	0	0	

Matrix (9.22) contains the three Equations (9.7) (9.8) and (9.9) and these may be solved using standard techniques of matrix algebra to give the solutions of Equations (9.10) and (9.11).

This technique is fairly trivial and is no more than an alternative method of solving the equations which arise through the use of the indicial method. A much more interesting matrix technique is associated with the Buckingham theorem. As was shown earlier, Buckingham's method consists in taking three of the variables associated with the problem and using these as repeating variables. Bearing in mind the method used earlier, it would be possible to set up the dimensional matrix in the form:

	H	g	ρ	Q	θ	μ	
L	1	1	-3	3	0	-1	
M	0	0	1	0	0	1	(9.23)
T	0	-2	0	-1	0	-1	

A solution can now be forced by rewriting the matrix, not in terms of the dimensions L, M and T, but in terms of the dimensions of the repeating variables H, g and ρ. Thus:

	H	g	ρ	Q	θ	μ
H	1	0	0	5/2	0	3/2
g	0	1	0	1/2	0	1/2
ρ	0	0	1	0	0	1

$$(9.24)$$

Each column to the right of the unit matrix now gives one π term. However, a much simpler approach is available.

Matrix (9.23) has been split into two parts which may be referred to conveniently as part (A) and part (B):

i.e. $A \quad\vdots\quad B$ (9.25)

Part (A) is the 3×3 matrix which must be transferred to the unit matrix in Matrix (9.24). This is done by inverting A and pre-multiplying, i.e.:

$$I = \text{unit matrix} = A^{-1} \cdot A \qquad (9.26)$$

Whatever was done to part (A) to give the unit matrix must also be done to part (B) to give the remainder of Matrix (9.24). Thus, if that remainder is called part (D):

$$A^{-1} \times B = D \qquad (9.27)$$

In full, this is:

$$\begin{bmatrix} 1 & 1 & -3 \\ 0 & 0 & 1 \\ 0 & -2 & 0 \end{bmatrix}^{-1} \begin{bmatrix} 3 & 0 & -1 \\ 1 & 0 & 1 \\ -1 & 0 & -1 \end{bmatrix} = \begin{bmatrix} 5/2 & 0 & 3/2 \\ 1/2 & 0 & 1/2 \\ 0 & 0 & 1 \end{bmatrix} \qquad (9.28)$$

Inversion and pre-multiplication of matrices are standard techniques for which computer programs are available (see *BASIC Matrix Methods*). This method is therefore amenable to a fully computerized solution which is particularly valuable when very complex problems involving large numbers of variables must be tackled.

9.5 Similarity theory

For two bodies to be geometrically similar, it is necessary that the ratio of corresponding lengths in corresponding positions be the same in both bodies. Kinematic similarity likewise requires that the ratio of corresponding velocities be the same and for physical or dynamic similarity between two systems it is necessary that the ratio of corresponding forces be the same in both systems. Generally speaking, geometric and dynamic similarity are prerequisites to kinematic similarity because similar velocities will be achieved only if the boundaries and accelerations are similar.

Because inertial forces are present in all accelerative fluid systems, it is convenient to use the inertial force as a basis for comparison. Thus, physical similarity between a model and prototype require that:

$$\left(\frac{F_i}{F}\right)_m = \left(\frac{F_i}{F}\right)_p \tag{9.29}$$

where F = any force and F_i = inertial force.

But $F_i \propto$ mass × acceleration $\tag{9.30}$

Mass $\alpha \ \rho l^3$ and acceleration $\propto V^2/l^2$.

Therefore: $F_i \propto \rho l^2 V^2 \tag{9.31}$

This short analysis is based on the idea that in two geometrically similar systems, all lengths are proportional to one standard length so that volume is proportional to l^3. Similarly, in kinematically similar systems, all velocities are proportional to one standard velocity.

Having evaluated the inertial force, other forces may be similarly determined. For example, the gravitational force is given by:

$$F_g \ \alpha \ \text{mass} \times g \propto \rho l^3 g \tag{9.32}$$

Thus

$$\frac{F_i}{F_g} \propto \frac{V^2}{lg}$$

and, from Equation (9.29), the requirement for dynamic similarity is:

$$\left(\frac{V^2}{lg}\right)_m = \left(\frac{V^2}{lg}\right)_p \tag{9.33}$$

Note here that the proportionality sign has been changed to an equals sign and this is possible because in similar systems, the coefficient of proportionality will have the same value in model and prototype. The square root of the ratio given in Equation (9.33) is known as the Froude number and is a standard for similarity of fluid systems involving gravitational forces. Similar standard numbers may be developed for other force actions. For example, if viscosity or viscous forces are important, it can be shown that:

$$F_v \propto \mu V l \tag{9.34}$$

Then

$$\frac{F_i}{F_v} = \left(\frac{\rho V l}{\mu}\right)_m = \left(\frac{\rho V l}{\mu}\right)_p \tag{9.35}$$

This ratio is known as the Reynolds Number.

Other standard numbers may be developed in a similar fashion. Some of these are shown in Table 9.2.

9.6 Application to models

Many models are designed and operated using the standard numbers developed in the previous section but complex models may require a full dimensional analysis. In either case, the general basis is to ensure equality of the standard number or, in the case of dimensional analysis, the dimensionless parameter in both model and prototype. For example, consider a model in which gravitational forces are important. From Equation (9.32) it can be seen that the velocity scale must be given by:

$$\frac{V_m}{V_p} = \left(\frac{l_m}{l_p}\right)^{1/2} \tag{9.36}$$

where $g = $ constant.

Discharge is given by the product of velocity and cross sectional area, which in similar systems is proportional to the square of the length, so that:

$$\frac{Q_m/l_m^2}{Q_p/l_p^2} = \left(\frac{l_m}{l_p}\right)^{1/2} \tag{9.37}$$

or

$$\frac{Q_m}{Q_p} = \left(\frac{l_m}{l_p}\right)^{5/2} \tag{9.38}$$

The same scale may be developed from the first parameter of Equation (9.14) and it is instructive to note that both of these are identical to Equation (8.9).

The force relationship is given by Equation (9.31) and substituting for V in Equation (9.36) then leads to the force scale:

$$\frac{F_m}{F_p} = \left(\frac{l_m}{l_p}\right)^3 \tag{9.39}$$

These scales have been developed for a Froudian model in which gravitational forces are important. With the model built to give geometric similarity the discharge would be set according to Equation

(9.38). This would ensure dynamic and kinematic similarity. Velocities and forces measured in the model could then be scaled up to prototype size using Equations (9.36) and (9.39).

Other scales may be developed in a similar manner. Model problems are often much more complex because more than one force might be important. In that case, each of the relevant parameters or standard numbers must have the same value in model and prototype. In many cases this is not possible and it is important then to choose the dominant force action and leave the others to be out of scale. This results in scale errors which, if the dominant force action is chosen correctly, will usually be small.

WORKED EXAMPLES

Example 9.1: SHIP: drag force on a ship model

A ship having a wetted surface area of $40 \, \text{m}^2$ is to be driven at $15 \, \text{m/s}$ in sea water. The skin friction is given by the formula $F = 0.55 \, AU^{1.85}$ where $A(\text{m}^2)$ and U (m/s) are the area and velocity of the full size ship. The corresponding formula for a $1/20$ scale model is $F = 0.75 AU^{1.95}$. Write a program to calculate the total resistance of the full size ship if the model resistance (at design velocity) was found to be $16.3 \, \text{N}$.

```
10 REM "SHIP"
20 CLS: KEY OFF
30 PRINT"This program will scale up ship model data."
40 PRINT
50 INPUT"Please enter the area of the ship (sq m)  "; AP
60 INPUT"Please enter the ship velocity (m/s)    ";UP
70 INPUT"Please enter the total model force (N)    ";TM
80 AM=AP/(20*20)
90 UM=UP/SQR(20)
100 FM= .75*AM*(UM ^ 1.95)
110 RM = TM - FM
120 RP = RM *(20^3)*(10060/9810)
130 FP = .55 * AP * (UP^1.85)
140 TOT = FP+RP
150 PRINT:PRINT"Total prototype force (N) = ";TOT
160 PRINT:PRINT:PRINT"program ended..."
170 END


This program will scale up ship model data.

Please enter the area of the ship (sq m)  ? 400
Please enter the ship velocity (m/s)    ? 15
Please enter the total model force (N)    ? 16.3

Total prototype force (N) =  101542.7

program ended...
Ok
```

Program notes

(1) Lines 10–40 clear screen, keys off, print headings.
(2) Lines 50–70 enter data.
(3) Lines 80–90 calculation of model area and velocity.
(4) Line 100 calculation of model skin friction force *FM*.
(5) Line 110 calculation of model gravity force = total force-*FM*.
(6) Line 120 scale up of model gravity force to prototype.
(7) Line 130 calculation of prototype skin friction force.
(8) Line 140 total prototype force = gravity + skin friction.
(9) Lines 150–170 print results and stop.

Example 9.2: RIVS: distorted river model

A distorted river model has its scales limited by the facilities available (length of tank, size of pumps etc.) and by the requirement that the model must be turbulent and feasible. Develop a program to assist in the choice of model scales. Pay attention to turbulence, roughness, discharge requirements and to velocity and depth measurement.

```
10 REM "RIVS"
20 CLS:KEY OFF
30 PRINT:PRINT"   This program assists the development"
35 PRINT"of scales for a distorted river model. Various"
40 PRINT"prototype data are required"
45 A=3
50 WHILE A<>1 AND A<>2
55 PRINT
60   PRINT"Choose the dynamic viscosity (nu) :"
65   PRINT
70   PRINT"  1) .000001  m2/sec (metric results)"
80   PRINT"       2) .0000107 ft2/sec (FPS results)"
90   PRINT:PRINT"1 or 2   ";
100  INPUT A
110  IF A= 1 THEN NU = .000001
120  IF A= 2 THEN NU = .0000107
130 WEND
140 PRINT:INPUT"Enter prototype velocity  "; V
150 PRINT:INPUT"Enter prototype discharge "; Q
160 PRINT:INPUT"Enter prototype depth   ";D
170 PRINT:INPUT"Enter prototype roughness   "; R
180 PRINT:INPUT"Enter hor scale (1/X),  X =  ";X
190 PRINT:INPUT"Enter vert scale (1/Y),  Y =  "; Y
200 CLS
210 PRINT :PRINT "Model Values:"
220 PRINT"--------------"
230 PRINT"Depth       = ";(D/Y)
240 PRINT"Velocity    = ";V/(Y^.5)
250 PRINT"Discharge   = ";Q/(X*(Y^1.5))
260 PRINT"Reynolds No = ";(V*D)/(NU*(Y^1.5))
270 PRINT"Roughness n = ";(R/(Y^(1/6)))*((X/Y)^.5)
280 PRINT:PRINT
```

```
290 PRINT"Which do you want to do:"
300 PRINT"    1) Redo with new hor and vert scales.
310 PRINT     "    2) Redo with new vert scale.
320 PRINT     "    3) Quit for now.
330 PRINT:INPUT ANS
340 IF ANS =3 THEN 380
350 IF ANS= 2 THEN GOTO 190
360 IF ANS = 1 THEN GOTO 180
370 GOTO 290
380 PRINT:PRINT"Program Ended..."
390 END
```

```
    This program assists the development
of scales for a distorted river model. Various
prototype data are required

Choose the dynamic viscosity (nu) :

    1) .000001  m2/sec (metric results)
    2) .0000107 ft2/sec (FPS results)

1 or 2   ? 1

Enter prototype velocity  ? .75

Enter prototype discharge ? 200

Enter prototype depth   ? 1.2

Enter prototype roughness  ? 0.015

Enter hor scale (1/X),  X =  ? 85

Enter vert scale (1/Y),  Y =  ? 60




Model Values:
--------------
Depth          =  .02
Velocity       =  .0968246
Discharge      =  5.062729E-03
Reynolds No    =  1936.494
Roughness n    =  9.023324E-03


Which do you want to do:
    1) Redo with new hor and vert scales.
    2) Redo with new vert scale.
    3) Quit for now.

? 3

Program Ended...
Ok
```

Program notes

(1) Lines 10–40 clear screen, keys off, print headings.
(2) Lines 50–130 loop to choose units and repeat question if incorrect entry.
(3) Lines 140–190 enter data for prototype and scales.
(4) Lines 200–270 calculate and print model data.
(5) Lines 280–370 routine to repeat calculation or stop.
(6) Lines 380–390 stop execution.

Example 9.3: MATR:

Write a program to undertake a dimensional analysis using matrix operations described in Section 9.4. Test it by operating on the Matrix (9.23).

```
100 REM MATRIX CALCS ON VAX BASIC ONLY
110 PRINT"THIS PROGRAM READS MATRICES A AND B FROM"
120 PRINT"DATA STATEMENTS AND CREATES MATRIX I, THE"
125 PRINT"INVERSE OF A, AND MATRIX D, THE PRODUCT"
130 PRINT"OF MATRICES I AND B."
140 MAT READ A(3,3)
150 MAT READ B(3,3)
160 MAT C=INV(A)
170 MAT D=C*B
180 MAT I=C*A
190 PRINT"MATRIX A :"
200 MAT PRINT A;
210 PRINT"MATRIX I :"
220 MAT PRINT I;
230 PRINT"MATRIX B :"
240 MAT PRINT B;
250 PRINT "MATRIX D :"
260 MAT PRINT D;
270 DATA 1,1,-3
280 DATA 0,0,1
290 DATA 0,-2,0
300 REM END OF MATRIX A
310 DATA 3,0,-1
320 DATA 0,0,1
330 DATA -1,0,-1
340 REM END OF MATRIX B
350 END


RUN
THIS PROGRAM READS MATRICES A AND B FROM
DATA STATEMENTS AND CREATES MATRIX I, THE
INVERSE OF A, AND MATRIX D, THE PRODUCT
OF MATRICES I AND B.

MATRIX A  :
1   1  -3
0   0   1
0  -2   0
```

```
MATRIX I :
1   1   0
0   1   0
0   0   1
  MATRIX B :
   3   0  -1
   0   0   1
  -1   0  -1
  MATRIX D :
   2.5   0   1.5
   0.5   0   0.5
   0     0   1
```

Program notes

(1) Lines 100–130 print headings.
(2) Lines 140–150 read matrix data.
(3) Lines 160–180 calculation of inverse and new matrices.
(4) Lines 190–260 print matrices.
(5) Lines 270–340 data for input matrices.
(6) Note that this form of BASIC is usually not available on micro computers. This program was run on VAX.

Example 9.4: PUTU: pump, turbine models

Develop a program to assist in the choice of model scales for pumps or turbines. Use it to determine a range of speeds, discharge and power in a model turbine for a given range of heads and a variety of scales.

Dimensional analysis yields characteristic equations which are usually written as:

$$Pumps \quad \left\{ \frac{(gH)^{1/2}}{ND}, \left[\frac{Q}{ND^3} \text{ or } \frac{P}{N^3 \rho D^5} \right] \right\} = 0$$

$$Turbines \quad \left\{ \frac{(gH)^{1/2}}{ND}, \left[\frac{Q}{D^2 (gH)^{1/2}}, \frac{P}{\rho D^2 (gH)^{3/2}} \right] \right\} = 0$$

```
10 REM "PUTU"
20 CLS: KEY OFF
30 PRINT"  This program will assist in the choice"
35 PRINT"of scales for model pumps or turbines."
40 PRINT"Use it to determine a range of speeds,"
45 PRINT"discharges and powers in a model turbine"
50 PRINT"for a given range of heads and a variety "
55 PRINT"of scales"
70 PRINT
80 INPUT "Enter prototype head (m)"; HP
90 INPUT"Please input prototype power (kw)   ";PP
```

```
100 INPUT "Enter prototype discharge (m3/s)";  QP
110 INPUT "Enter prototype speed (rev/min) ";NP
120 INPUT "Enter prototype density "; RHOP
130 INPUT "Enter scale (1/X)     X = ";X
140 INPUT "Enter model operating head   (m)";HM
150 INPUT "Enter model density ";RHOM
160 PRINT
170 CHOICE =3
180 WHILE CHOICE<>1 AND CHOICE<>2
190     PRINT "1.    Turbine"
200     PRINT "2.    Pump"
210     INPUT "Enter 1 or 2 ";CHOICE
220 WEND
230 ON CHOICE GOTO 250,350
240 REM TURBINE CALCULATIONS
250 PRINT "Turbine"
260 PRINT
270 MS = NP*(HM/HP)^.5*X
280 PRINT "Model speed (rev/min) = ";MS
290 MP = PP*(RHOM/RHOP)/(X*X)*(HM/HP)^1.5
300 PRINT "Model power (kw) = ";MP
310 MD = QP*(1/X)^2*(HM/HP)^.5
320 PRINT "Model discharge (m3/s)= ";MD
330 GOTO 430
340 REM PUMP CALCULATIONS
350 PRINT "Pump"
360 PRINT
370 NM = NP*(HM/HP)^.5*X
380 PRINT "Model speed (rev/min) = ";NM
390 MD = QP*(NM/NP)*(1/X)^3
400 PRINT "Model discharge (m3/s) = ";MD
410 MP = PP*(NM/NP)^3*(1/X)^5
420 PRINT "Model power (kw) = ";MP
430 PRINT "Another calculation ? (Y or N) ";
440 INPUT A$
450 IF A$ = "y" OR A$ = "Y" THEN GOTO   70
460 PRINT "Program ended ...... "
470 END
```

```
    This program will assist in the choice
of scales for model pumps or turbines.
Use it to determine a range of speeds,
discharges and powers in a model turbine
for a given range of heads and a variety
of scales
```

```
Enter prototype head (m)? 44
Please input prototype power (kw) ? 1700
Enter prototype discharge (m3/s)? 3.85
Enter prototype speed (rev/min) ? 302
Enter prototype density ? 1000
Enter scale (1/X)     X = ? 4
Enter model operating head   (m)? 7.5
Enter  model density ? 1000
```

```
1.    Turbine
2.    Pump
Enter 1 or 2 ? 1
Turbine

Model speed (rev/min) =   498.7366
Model power (kw) =   7.477249
Model discharge (m3/s)=   9.934478E-02
Another calculation  ? (Y or N) ? n
Program ended ......
Ok
```

Program notes

(1) Lines 10–70 clear screen, keys off, print headings.
(2) Lines 80–160 enter data.
(3) Lines 170–220 choose pump or turbine only.
(4) Line 230 route to turbine routine or pump routine.
(5) Lines 240–320 calculation of model values from turbine equations.
(6) Lines 340–420 calculation of model values from pump equations.
(7) Lines 430–470 provide opportunity to repeat calculation or stop.

Table 9.1. Dimensions of fluid variables

Variable	Symbol	Dimensions
Length	L	$[L]$
Time	T	$[T]$
Mass	M	$[M]$
Velocity	V	$[L]/[T]$
Acceleration	a	$[L]/[T]^2$
Discharge	Q	$[L]^3/[T]$
Kinematic Viscosity	ν	$[L]^2/[T]$
Force	F	$[M][L]/[T]^2$
Mass Density	ρ	$[M]/[L]^3$
Specific Weight	γ	$[M]/([L]^2[T]^2)$
Pressure; shear stress	$p; \tau$	$[M]/([L][T]^2)$
Dynamic viscosity	μ	$[M]/([L][T])$
Surface tension	σ	$[M]/[T]^2$
Bulk modulus	K	$[M]/([L][T]^2)$

Table 9.2. Standard dimensionless numbers

Name	Format	Comment
Froude number	$V/(gL)^{1/2}$	May also be defined as the square of this quantity. Relevant to gravity forces.
Reynolds number	VL/v	Relevant to viscous force action.
Weber number	$V(\rho L/\sigma)^{1/2}$	May also be defined as the square of this quantity. Refers to the action of surface tension forces.
Euler number	$\Delta p/(\rho V^2)$	Also expressed as $V/(2\Delta p/\rho)^{1/2}$. Important when pressure forces exist. Δp = pressure difference.
Cauchy number	$\rho V^2/K$	Relevant when compressibility is important.
Mach number	V/C	Refers to compressibility effects and high-speed flow. C = local velocity of sound wave in fluid.

PROBLEMS

(9.1) A spherical balloon that is to be used in air at 150 °C is tested using a 1/10 scale model and later, as a further check by a 1/3 scale model. Both models were towed at different speeds in a deep water tank. The results of the tenth scale tests were:

Drag (N)	12.15	22.1	57.05	72.3	97.8
model speed (m/s)	0.5 5	0.7	1.1	1.2	1.5

Write a program to predict the results of the 1/3 scale tests and to determine the full scale drag on the balloon.

(9.2) Write a program to determine velocity, discharge, time and force scales for models designed to simulate gravitational forces and for models designed to simulate viscous forces. Allow the user to choose the type of model. Incorporate routines for determining model discharge and for predicting prototype velocities and forces based on measurements made in the model.

(9.3) A 1/200 scale model of an office building is tested in a wind tunnel to determine pressures and forces on the full scale structure. The model is tested in a wind tunnel at 20 m/s at 20° C and information is required for wind speeds of 100 km/h, 120 km/h and 150 km/h at 10 °C on the full scale structure. Pressures and velocities are correlated by $P = 0.5\, Cp\, \rho V^2$ where p = pressure difference, Cp = pressure coefficient, ρ = density and V = velocity. At key points on the model pressure coefficients were found to be $+1.0$, $+1.2$, -2.7, -0.8 and -1.3. Determine the pressures which would occur on the full size structure at the three design wind speeds.

(9.4) In models of river estuaries the rise and fall of the tide at the downstream end must be scaled both for water level and for time. Write a program to read and record prototype data giving level as a function of time. Convert this to model data and print a table showing model elevation as a function of model time. Permit different scales to be considered and include the possibility of distorted models with different horizontal and vertical scales.

(9.5) A $1/10.9$ scale model is constructed of the condenser pipework complex in a thermal power plant. Air is used to simulate the water flow and velocity distribution measurements are made at all bends to obtain a measure of the flow distribution across the 2-m wide rectangular ducts. Write a program to record experimental results and transform them to full scale values.

(9.6) The program PUTU is not well written because; (a) it permits independent specification of head, discharge and power, (b) it uses different equations for pumps and turbines when there is no need to do so (enter the same values for a pump and a turbine to check), and (c) it contains a number of GOTO statements which would be better replaced with WHILE-WEND loops.

Rewrite the program to take care of these criticisms. Replace all GOTOs. Use one set of equations for pumps and turbines and permit the user to specify any two, but only two, of head power and discharge.

Index